T0100761

Once Upon a Prime

Once Upon a Prime

The Wondrous Connections Between Mathematics and Literature

Sarah Hart

Printed in the United States of America. For information, address Flatiron Books, 120 Broadway, New York, NY 10271.

www.flatironbooks.com

Designed by Donna Sinisgalli Noetzel

The Library of Congress Cataloging-in-Publication Data is available upon request.

ISBN 978-1-250-85088-1 (hardcover)
ISBN 978-1-250-85089-8 (ebook)

Our books may be purchased in bulk for promotional, educational, or business use. Please contact your local bookseller or the Macmillan Corporate and Premium Sales Department at 1-800-221-7945, extension 5442, or by email at MacmillanSpecialMarkets@macmillan.com.

First Edition: 2023

10 9 8 7 6 5 4 3 2 1

For Mark, Millie, and Emma

Contents

Part III: Mathematics Becomes the Story

Once Upon a Prime

Introduction

"Call me Ishmael." This has to be one of the most famous opening sentences in literature. I'm embarrassed to say that I didn't get past it for a long time—*Moby-Dick* was in the guilt-inducing category of "books you should have read," which obviously made me rebel against doing so, as I feared it would be the worst of all things: worthy. Thank goodness I decided one day to finally take the plunge, because it's probably not an exaggeration to say it changed my life. It set me thinking about the links between mathematics and literature, which led ultimately to this book.

It all started when I heard a mathematician mention that *Moby-Dick* contains a reference to cycloids. The cycloid is a beautiful mathematical curve—the mathematician Blaise Pascal found it so distractingly fascinating that he recounted thinking about it to relieve the pain of toothache. But applications to whaling are not usually listed on its résumé. Intrigued, I decided it was high time to finally read this Great American Novel. To my surprise and delight, I found that right from the start, *Moby-Dick* abounded with mathematical metaphors. The more Melville I read, the more mathematics I discovered. And it wasn't just Melville. Leo Tolstoy writes about calculus, James Joyce about geometry. Mathematicians appear in work by authors as disparate as Arthur Conan

Doyle and Chimamanda Ngozi Adichie. And how about the fractal structure that underlies Michael Crichton's *Jurassic Park* or the algebraic principles governing various forms of poetry? Mathematical references in literary works go back at least as far as Aristophanes's *The Birds*, first performed in 414 BCE.

There have been occasional academic studies on mathematical aspects of specific genres or authors. But even in the case of Melville, whose affinity for mathematics is (for me) so obvious in his work, I could find just a handful of scholarly articles. The more holistic connections between mathematics and literature have not received the attention they deserve. My goal in this book is to convince you not only that mathematics and literature are inextricably, and fundamentally, linked, but that understanding these links can enhance your enjoyment of both.

Mathematics is often viewed as being quite separate from literature and other creative arts. But the perceived boundary between them is a very recent idea. For most of history, mathematics was part of every educated person's cultural awareness. More than two thousand years ago, Plato's *Republic* put forward the ideal curriculum of arts to be studied, which medieval authors split into the trivium (grammar, rhetoric, logic) and the quadrivium (arithmetic, music, geometry, astronomy). Together, these are the essential liberal arts. There is no artificial dichotomy here between "mathematics" and "art."

The eleventh-century Persian scholar Omar Khayyám, to whom the poetry collection known as the *Rubaiyat* is attributed (modern scholars believe it to be the work of several authors), was also a mathematician, creating beautiful geometrical solutions to mathematical problems whose full algebraic solutions would not be found for another four hundred years. In the fourteenth century, Chaucer wrote both *The Canterbury Tales* and a treatise on the astrolabe. There are innumerable such examples, not least that of Lewis Carroll, who, of course, was a mathematician first and author second.

But there is a deeper reason why we find mathematics at the heart of literature. The universe is full of underlying structure, pattern, and regularity, and mathematics is the best tool we have for understanding it—that's why mathematics is often called the language of the universe, and why it is so vital to science. Since we humans are part of the universe, it is only natural that our forms of creative expression, literature among them, will also manifest an inclination for pattern and structure. Mathematics, then, is the key to an entirely different perspective on literature. As a mathematician, I can help you see it too.

I've always loved patterns—whether that's patterns of words, numbers, or shapes. I've loved patterns since before I knew that what I was doing was mathematics. Gradually it became clear that I was going to be a mathematician, but that came with consequences. In the British education system in recent decades, mathematics has come to be treated exclusively as a science subject, far removed from the humanities. If you want to study mathematics after the age of sixteen, you probably have to pick the "science" stream. At the end of my very last English class at school, in 1991, the teacher gave me a lovely handwritten note with a long list of books she thought I might like, saying, "Sorry to lose you to the lab." I was sorry to be considered lost, too. But I wasn't lost—and if you've ever had to "choose" one subject over another, you aren't lost either. I love language; I love the way words fit together; I love the way that fiction, like mathematics, can create, play with, and test the limits of imaginary worlds. I went off to Oxford to study mathematics, very happy to be living one street away from the pub where my childhood literary heroes C. S. Lewis and J.R.R. Tolkien had met each week to discuss their work.

After completing a master's degree and Ph.D. in Manchester, in the north of England, I moved to London for a job at Birkbeck, one of the colleges of the University of London, in 2004, and became a full professor there in 2013. During all this time, although my "day job," so to speak, has been teaching and research, mainly in the area of abstract algebra known as group theory, I became increasingly interested in the history of mathematics, particularly in how mathematics is part of our broader cultural experience. I've always felt that what I do as a mathematician fits in with other creative arts, like literature or music. Good mathematics, like good writing, involves an inherent appreciation of structure, rhythm, and pattern. That feeling we get when we read a great novel or a perfect sonnet—that here is a beautiful thing, with all the component parts fitting together perfectly in a harmonious whole—is the same feeling a mathematician experiences when reading a beautiful proof. The mathematician G. H. Hardy wrote that "a mathematician, like a painter or poet, is a maker of patterns. . . . The mathematician's patterns, like the painter's or the poet's, must be beautiful; the ideas, like the colours or the words, must fit together in a harmonious way. Beauty is the first test: there is no permanent place in the world for ugly mathematics."

Becoming Gresham Professor of Geometry in 2020 gave me the chance to bring together my decades of thinking about mathematics and its place in history and culture. This professorship is one of the few Tudor jobs still around—it was created in 1597 in the will of the Elizabethan courtier and financier Sir Thomas Gresham, and I'm the thirty-third person, and the first woman, to do it. I get to give public lectures on any mathematical subject of my choosing, though fortunately it's been well over a century since professors were required to deliver each lecture twice: once in English and once in Latin.

So with being a professor of mathematics at Birkbeck and concurrently a professor of geometry at Gresham, as well as raising two wonderful daughters, I know what you are thinking: *Sarah, what do you do with all your spare time?* The answer is that, as I've always done, I read. Constantly and widely. The best thing about e-readers is that there are no pages to turn, which meant I could read even with a sleeping baby in my arms. That's how I finally found the time to read *War and Peace*, which was full of mathematical surprises.

Each year my good friend Rachel and I set ourselves the challenge of reading the Booker Prize shortlist before the winner is announced. This gives us about six weeks to read six books. In 2013, one of the shortlisted books (and in fact the eventual winner) was Eleanor Catton's *The Luminaries*. Catton made use of several structural constraints in the novel, including a mathematical sequence known as a geometric progression. There are hidden clues and rewards for the reader aware of the mathematics behind the scenes—a haul of stolen gold worth precisely £4,096 is not a coincidence, for example—and understanding the geometric progression unfolding throughout gives you another dimension of enjoyment. This is just one of many literary uses of mathematical structures that I'll show you in this book.

It's worth pointing out as well that the links between mathematics and literature do not run in just one direction. Mathematics itself has a rich heritage of linguistic creativity. Going back to early India, Sanskrit mathematics followed an oral tradition. Mathematical algorithms were encoded in poetry so that they could be passed on by word of mouth. We think of mathematical concepts as relating to precise, fixed words: square, circle. But in the Sanskrit tradition, your words must fit into the meter of your poem. Number words, for example, can be replaced with words for relevant objects. The number 1 can be represented by anything that is unique, like the moon or the earth, while "hand" can mean 2, because we have two hands—but so can "black and white," because they form a pair. An expression like "three voids teeth" doesn't mean a visit to the dentist, but that three zeros should follow the number of teeth we have: a poetic way to say 32,000. The huge array of different words and meanings lends a compelling richness to the mathematics.

Mathematical language continues to be figurative—when we need new words for things, we reach for metaphors. Once these words have been established for long enough, we tend to forget that they have other layers of meaning. But sometimes circumstances can intervene to remind us. As a master's student, I spent a semester studying at the University of Bordeaux, in southwest France, and reading mathematics in French lent a slight air of surreality to the mathematics because of the use of words and metaphors I didn't yet know in a mathematical context. Those few months of study opened my eyes forever to the creative metaphorical language underpinning much of mathematics. Learning a subject called algebraic geometry in French, I got a distinctly agricultural vibe from the word *gerbe*, which until then I had previously been aware of only in the phrase *gerbe de blé* (wheat sheaf). Sometimes you can overtranslate—for a while I thought there was a result called the walrus theorem, because the French word *morse* translates to "walrus," when in fact it was named after its discoverer, the respected mathematician (and non-walrus) Marston Morse.

Just as mathematics makes use of literary metaphors, literature abounds with ideas that a mathematically attuned eye can detect and explore. This adds an extra dimension to our appreciation of a work of fiction. Melville's cycloid, for example, is a curious curve with many wonderful properties, but unlike curves such as the parabola and ellipse, you probably haven't heard of it unless you are a mathematician. That's a real shame, because the properties of this curve are so beautiful that it was nicknamed "the Helen of geometry." Making a cycloid is quite easy. Imagine a wheel rolling along a flat road. Now mark a point on the rim somehow, say with a blob of paint. That blob will trace out a path in space as the wheel rolls, and this path is called a cycloid. This is a fairly natural idea, but we don't have evidence of its being studied until the sixteenth century, and things didn't really heat up until the seventeenth and eighteenth centuries, when it seemed that everyone who was interested in mathematics had something to say. It was Galileo, for example, who came up with the name "cycloid": he wrote that he had worked on cycloids for fifty years.

So the fact that the cycloid gets a mention not just in *Moby-Dick* but in two great works of eighteenth-century literature, *Gulliver's Travels* and *Tristram Shandy*, again shows us mathematics in its rightful place—not "other" but part of intellectual life. When Gulliver visits the land of Laputa, he finds the inhabitants obsessed with mathematics. Dining with the king, he reports that "servants cut our bread into cones, cylinders, parallelograms, and several other mathematical figures." There is a shoulder of mutton "cut into an equilateral triangle" and "a pudding into a cycloid." Meanwhile, over at Shandy Hall, Tristram's uncle Toby is having terrible trouble trying to construct a model bridge. After consulting various learned sources (there's even a reference to a real-life mathematical paper in the extremely clever-sounding journal *Acta Eruditorum*), he decides, rather rashly, that a cycloid-shaped bridge is the way forward. But it doesn't go well: "My uncle Toby understood the nature of a parabola as well as any man in England—but was not quite such a master of the cycloid;—he talked however about it every day—the bridge went not forwards."

Part of the enjoyment of reading *Tristram Shandy* and other great books is the dazzling richness and breadth of their allusions—literary, cultural, and, yes, mathematical. If you're reading classic literature, then it makes sense to be at least a little familiar with works of Shakespeare because of their profound literary and cultural influence. Is there a mathematical equivalent to the works of Shakespeare, references to which abound in classic literature? A strong contender would be the books of Euclid, known collectively as *The Elements of Geometry*, or just Euclid's *Elements*. They are probably the most influential mathematics books of all time.

There's an anecdote about how the philosopher Thomas Hobbes got hooked on geometry, told by his biographer John Aubrey:

> *Being in a gentleman's library Euclid's* Elements *lay open, and 'twas the forty-seventh proposition in the first book. He read the proposition. "By G—d," said he, "this is impossible!" So he reads the demonstration of it, which referred him back to such a proof; which referred him back to another, which he also read. . . . At last he was demonstratively convinced of that truth. This made him in love with geometry.*

This is a nice story, and it tells us a lot about how mathematics was viewed. Euclid's *Elements* lay open, notice, because Hobbes was in "a gentleman's library," not "a mathematician's study." This stuff was considered part of the well-rounded education of an informed person. More than this, Aubrey assumes that we, the readers, are familiar with Euclid.

He refers to Book I, Proposition 47, as if we will know it. We do know it, in fact, because it is Pythagoras's theorem.

The beautiful certainties encapsulated in Euclidean geometry—axioms and definitions leading inexorably to theorems and proofs—have both inspired and consoled literary figures, from George Eliot and James Joyce, both in their different ways lovers of mathematics, whom we'll meet in Chapter 6, to poets like William Wordsworth and Edna St. Vincent Millay. In his "Prelude," Wordsworth speaks of geometry bringing "a pleasure quiet and profound" that can "beguile [your] sorrow":

Mighty is the charm
Of those abstractions to a mind beset
With images, and haunted by herself,
And specially delightful unto me
Was that clear synthesis built up aloft
So gracefully; . . .
. . . an independent world
Created out of pure intelligence.

Everybody knew about the perfection of Euclid, so in the nineteenth century the tremendously exciting discovery of geometries beyond the Euclidean world—things like so-called non-Euclidean geometries, where parallel lines can sometimes meet—instantly caught the public's imagination. I'll show you how these ideas have been interpreted in literature by everyone from Oscar Wilde to Kurt Vonnegut. By seeing mathematics and literature as complementary parts of the same quest to understand human life and our place in the universe, we immeasurably enrich both fields.

In Part I of this book, we will explore the fundamental structures of literary texts, from plot in novels to rhyme schemes in verse. I'll show you the underlying mathematics of poetry. I'll give you the lowdown on writing that, like *The Luminaries*, deliberately uses constraints, such as the mathematically inspired work of the French literary group the Oulipo, whose members included Georges Perec and Italo Calvino. In the house of literature, these are the foundations, the load-bearing beams. It's here that we'll find mathematical ideas hidden in plain sight.

What comes next is the decoration, the wallpaper, the carpets. Many authors have reached for mathematical metaphors in their writing, and the symbolism of numbers is rich and ancient. These turns of phrase, metaphors, and allusions will be our focus in Part II of the book.

But who lives in our house? What is our writing about? In Part III, I'll show you how mathematics can become part of the story—with novels featuring overtly mathematical themes and sometimes even mathematicians as characters. We will look at mathematical ideas that have caught the public's imagination, from fractals to the fourth dimension, and how they have been explored in fiction. We'll look too at how stereotypes of mathematicians, and the idea of mathematics itself, have been used in fiction.

If you don't yet love mathematics, I want this book to show you the beauty and wonder of it, how it is a natural part of our creative lives, and why it deserves its place with literature in the pantheon of the arts. I want it to give you an extra perspective on the writing and writers you know, introduce you to writing you don't, and give you a new way of experiencing the written word. If you happen to be a mathematician, then you already have poetry in your soul, but we'll look at how that is manifested in places you may never have realized, as part of an enduring conversation between literature and mathematics. I warn you: you're going to need a bigger bookcase.

Part I

Mathematical Structure, Creativity, and Constraint

1

One, Two, Buckle My Shoe

The Patterns of Poetry

The connections between mathematics and poetry are profound. But they begin with something very simple: the reassuring rhythm of counting. The pattern of the numbers 1, 2, 3, 4, 5 appeals to young children as much as the rhymes we sing with them ("Once I caught a fish alive"). When we move on from nursery rhymes, we satisfy our yearning for structure in the rhyme schemes and meter of more sophisticated forms of poetry, from the rhythmic pulse of iambic pentameter to the complex structure of poetic forms like the sestina and the villanelle. The mathematics behind these and other forms of poetic constraint is deep and fascinating. I'll share it with you in this chapter.

Think of the nursery rhymes of your childhood. I bet you can still remember the words. That's the power of pattern—our mathematical brains delight in it. The subliminal counting of rhythm and rhyme feels so natural that it helps us remember, hence the oral tradition of poems telling the deeds of great heroes. Many traditional rhymes involve counting up cumulatively, adding a new line with each verse and counting back down to one every time. There's an old English folk song, "Green Grow the Rushes, O," which builds up to twelve—the last line of every verse is the melancholy "One is one and all alone and ever more shall be so." Meanwhile, the Hebrew *Echad Mi Yodea* ("Who Knows One") rhyme, traditionally sung on Passover, uses rhythm and counting to teach children important aspects of the Jewish faith. It ends with "four are the matriarchs, three are the patriarchs, two are the tablets of the covenant, One is our God, in heaven and on earth."

There are many mathematical mnemonics that we may have learned at school for remembering things like the first few digits of π. "How I wish I could calculate pi": that's not me expressing a desire to calculate π, it's the mnemonic. The number of letters in each word tells you the next number in the decimal, which begins 3.141592. If you need more digits, a longer mnemonic is "How I need a drink, alcoholic in nature, after the heavy lectures involving quantum mechanics!" That one has been around for at least a century and is credited to the English physicist James Jeans. In fact, it's now a niche hobby to compose verse in "pilish," in which the word lengths are defined by the digits of π.[1] My favorite example of this is "Near a Raven," a pilish version of Edgar Allan Poe's "The Raven," by Michael Keith:

Poe, E.

Near a Raven

Midnights so dreary, tired and weary.
Silently pondering volumes extolling all by-now obsolete lore.
During my rather long nap—the weirdest tap!
An ominous vibrating sound disturbing my chamber's
antedoor.
"This," I whispered quietly, "I ignore."

There's no need to learn this poem in its entirety, though—it's been estimated that a mere forty digits of π are enough to calculate the circumference of the entire known universe accurate to less than the size of a hydrogen atom. So the first verse alone is more than enough for all practical purposes.

The pilish "Raven" is based on a mathematical constant, but its contents aren't mathematical. There is, however, at least one well-known poem that poses a mathematical puzzle. You may know it:

As I was going to St. Ives,
I met a man with seven wives.
Each wife had seven sacks,
Each sack had seven cats,
Each cat had seven kits.
Kits, cats, sacks, and wives,
How many were going to St. Ives?

I remember trying to multiply all those sevens as a kid—only to realize I'd fallen for the oldest misdirection trick in the book.

Much more sophisticated mathematical problems have been expressed in verse, though. As I mentioned in the introduction, it was the standard format for mathematics in the Sanskrit tradition. The twelfth-century Indian mathematician and poet Bhaskara wrote all his mathematical works in verse. Here is one of the poems in a book he dedicated to his daughter Lilavati:

Out of a swarm of bees, one fifth part settled on a blossom of Kadamba,
and one third on a flower of Silindhri;
three times the difference of those numbers flew to the bloom of a Kutaja.
One bee, which remained, hovered and flew about in the air,
allured at the same moment by the pleasing fragrance of jasmine and pandanus.
Tell me, charming woman, the number of bees.

What a lovely way to write about algebra!

We don't tend to write our mathematics in verse nowadays, more's the pity, but the aesthetic link with poetry remains: the goal of both is beauty, a beauty that makes a virtue of economy of expression. Poets and mathematicians alike have praised each other's specialisms. "Euclid alone has looked on Beauty bare," wrote the American poet Edna St. Vincent Millay in a 1922 sonnet paying homage to Euclid's geometry. For the Irish mathematician William Rowan Hamilton, both mathematics and poetry can "lift the mind above the dull stir of Earth." Einstein is reported to have said that mathematics is the poetry of logical thought. A mathematical proof, for example, if it's any good, has a lot in common with a poem. In both cases, each word matters, there are no superfluous words, and the goal is to express an entire idea in a self-contained, usually fairly short, and fairly structured way.

I'm going to show you a proof now, because it's a beautiful thing and it is pure poetry. It's the proof, attributed to Euclid (though we don't really know who came up with it), that there are infinitely many prime numbers. Remember, primes are the numbers, like 2, 3, 5, 7, and so on, that can't be divided up into smaller whole number parts. The number 4, for instance, isn't prime because you can break it up as 2×2. And 6 is 2×3. Every one of the counting numbers after 1 is either a prime number or can be broken up (the technical term is "factorized") into prime numbers, and even more brilliantly, this can be done in really only one way, as long as you are happy to say that 2×3 is basically the same thing as 3×2. By the way, the number 1 feels as if it ought to be prime because it can't be divided up, but we exclude 1 from the list because otherwise you'd have to say that $6 = 1 \times 2 \times 3 = 1 \times 1 \times 2 \times 3 = 1 \times 1 \times 1 \times 2 \times 3 = \cdots$ and there would be infinitely many ways to factorize every number—yuck! We get around this by defining a prime number as a number greater than 1 whose only factors are 1 and itself.

Understanding the prime numbers is as important to math as understanding the chemical elements in science, because just as every chemical substance is made up of a precise combination of elements (every molecule of water, or H_2O, has exactly two hydrogen atoms and one oxygen atom, for instance), every whole number has a particular prime decomposition. One of the most exciting discoveries of early mathematics was that, unlike chemical elements, the prime numbers go on forever. Actually, at the time, the contrast would have been even more stark, because for the ancient Greeks there were just four elements—earth, air, fire, and water—that were believed to make up all things.

Here's a proof that there are infinitely many prime numbers:

What if we had a list of all primes, a finite list?
It would start with 2, then 3, then 5.
We could multiply all the primes together, and add 1 to make a new number.
The number is 2 times something plus 1, so 2 can't divide it.
The number is 3 times something plus 1, so 3 can't divide it.
The number is 5 times something plus 1, so 5 can't divide it.
None of the primes on our list can divide it.
Either our number is prime, or a new prime not on our list divides it.
Either way, the list isn't complete. It can't be done.
There can't be a finite number of primes.
QED

It's a poem, I tell you!

The resonances between poetry and mathematics were expressed well by the American poet Ezra Pound in *The Spirit of Romance* (1910): "Poetry is a sort of inspired mathematics, which gives us equations, not for abstract figures, triangles, spheres and the like, but equations for the human emotions." Pound made another analogy between mathematics and poetry—the way that both can be open to many layers of interpretation.[2] I would say that mathematicians have a very similar understanding of what makes the greatest mathematics: concepts that hold within them many possible interpretations—structures that can be found in different settings and so have a universality to them. The key thing here is that the elegant brevity of a mathematical expression, just like a poem, can encompass multiple layers of meaning, and the more layers and interpretations it can contain, the greater the artistry. Mathematics, like Walt Whitman, contains multitudes, both literally and allegorically. The only difference is that we hope it does not contradict itself!

It's quite hard to give a definition of what poetry is. Sometimes it rhymes, there are almost always line breaks, there's usually a rhythm, a meter, and so on. What we can broadly say is that poems have some sort of constraint, whether that's a meter (iambic pentameter, for example), or a rhyme scheme, or a given number of lines in each stanza. Even completely free verse will probably have line breaks, stanzas, and rhythm. One occasionally hears expressed that understanding how something is put together takes away the mystery and therefore spoils it. We don't want to know how the magician does his tricks—we want to believe in magic. The difference is that poetry is more than artifice. How can understanding something do anything other than add to your appreciation of it? That's how I feel about the underlying mathematics of structure and pattern.

Submitting yourself voluntarily to a particular constraint spurs creativity. The discipline required means you have to be inventive, creative, and thoughtful. In *haiku*, with their seventeen syllables, no syllable can be wasted. On a rather less exalted level, the humorous limerick form has to get from setup to payoff in just five lines. The Irish poet Paul Muldoon made the brilliant comment that poetic form "is a straightjacket in the sense that straightjackets were a straightjacket for Houdini." This may set the record for most uses of the word "straightjacket" in a sentence, but the sentiment is exactly right—the constraint itself is part of the genius of the work.

Constraints in poetry come in many flavors. In the Western tradition, particular rhyme schemes have been favored, and a handful of rhythms have been adopted—those iambs and trochees of classical verse. There is counting, pattern, and therefore mathematics behind both types of constraint. But in other traditions, different pattern-creating devices are used that involve more explicit use of numbers. That's where we'll begin our discussion of the mathematics of poetic constraints.

Let me tell you a story that begins in the imperial court of eleventh-century Japan. Murasaki Shikibu, a noblewoman at the court and lady-in-waiting to Empress Shoshi, wrote what is thought to be one of the very earliest novels, *The Tale of Genji*. An epic novel of courtly love and heroism, it is a Japanese classic, still read a millennium after it was written. One of the novel's distinctive features is characters' use of poetry in conversation, quoting or modifying well-known verses or saying the first parts of them (just as we might do when we say, for instance, "A stitch in time" rather than "A stitch in time saves nine"). Many of the poems in *The Tale of Genji* are in what is called the *tanka* form. This is one example of a more general style of classic Japanese poetry called *waka*. Like the more modern *haiku*, such poems feature lines of 5 and 7 syllables, but where *haiku* has a 5–7–5 pattern with 17 syllables in total, *tanka* has 5–7–5–7–7, for a total of 31 syllables. (In fact, what is counted are not exactly "syllables" but "sounds," a subtle but important distinction, which I beg experts in Japanese poetry to forgive me for not making in more detail.)[3]

For a mathematician, the connection with prime numbers is inescapable. Look at the *haiku*: 3 lines, lengths 5 and 7 syllables, and a total of 17 syllables. The numbers 3, 5, 7, and 17 are all prime numbers. With the *tanka*, there are 2 lines of 5 syllables and 3 lines of 7 syllables—and again, 2, 3, 5, 7, and 31 are all prime. Is this significant? I have read that the 5–7 pairing arose from an earlier "natural" 12-syllable entity, which is broken into two parts with a slight pause. Making the break at 5–7 certainly seems to me to be more exciting and dynamic than the dully exact 6–6 split or the too unbalanced 4–8, so perhaps that's how it came about. Since primes can't be divided further, the 5–7 break perhaps helps to categorize the lines as separate indivisible entities, whereas 4, 6, and 8 all have "fault lines" that would arguably weaken the structure.

Centuries after *The Tale of Genji* was written, a game became fashionable in the parlors of sixteenth-century Japanese aristocrats: *Genji-ko*. The hostess would secretly choose five incense sticks from a selection of different scents; some of the five scents might be the same. She would then burn them one after the other, and the guests would try to guess which scents were the same and which were different. So you might think that all the scents are different. Or perhaps the first and third scents are the same, and all the others are different. The various possibilities would be represented by little diagrams like this:

The far left diagram represents all scents being different; the next has just the first and third matching; in the next the first, third, and fifth match, as do the second and fourth; the far right diagram has the second, third, and fourth matching, as well as the first and fifth. To help people describe what their guess was, each of the different possibilities was named after a chapter from *The Tale of Genji*—it turns out there are fifty-two possibilities, from "all different" to "all the same" and everything in between.[4] Several editions of *The Tale of Genji* even featured these patterns next to the corresponding chapter headings. The patterns themselves took on a life of their own—they were used as heraldic crests and in kimono designs.

Meanwhile, thousands of miles away in Tudor England, George Puttenham included diagrams like this in his 1589 book *The Arte of English Poesie*:

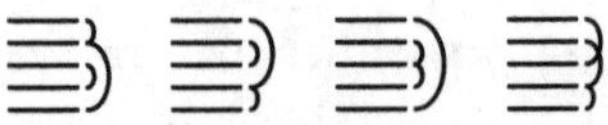

They look just like sideways versions of *Genji-ko* pictures! In particular, compare

and

What on earth is going on? Well, Puttenham is describing possible rhyme schemes in a five-line stanza, giving diagrams to aid the reader's comprehension (or as he put it, "I set you downe an occular example: because ye may the better conceive it").

The rhyme scheme of a poem, or of a stanza within a poem, is simply the pattern of rhymes in the last words of the lines. The earliest poems we encounter are songs and nursery rhymes with simple rhyme schemes:

Mary had a little lamb
Its fleece was white as snow
And everywhere that Mary went
The lamb was sure to go.

This is a four-line poem—a "quatrain"—with the rhyme scheme *abcb*, which means that the second and fourth lines rhyme with each other, but not with the remaining lines. By contrast, here's a quatrain from John Donne's poem "The Sun Rising":

Busy old fool, unruly sun,
Why dost thou thus,
Through windows, and through curtains call on us?
Must to thy motions lovers' seasons run?

This time, the scheme is *abba*.

If you ask a child to write you a poem, chances are you'll get a quatrain. As an experiment, I asked my daughter Emma just now to write me a poem "for Mummy's book." She came back three minutes later with this excellent mathematical verse:[5]

Endless numbers
You could count them till you die
It can outlive the universe
That is Pi.

I guess that could be either *abab* or *abcb*, depending on whether you think "numbers" rhymes with "universe."

For quatrains (four lines), there are fifteen potential rhyme schemes. From most to least rhymes, we have *aaaa* (boring), *aaab*, *aaba*, *aabb*, *abaa*, *abab*, *abba*, *abbb*, *aabc*, *abac*, *abbc*, *abca*, *abcb*, *abcc*, and *abcd* (not rhyming at all). Puttenham said that only three of these were allowable: even these he rather damns with faint praise. He describes *aabb* as "the most vulgar" (meaning commonplace), *abab* as "usuall and common," and, finally, *abba* as "not so common but pleasant and allowable inough." John Donne must be so relieved!

But enough with the quatrains. For a five-line poem, which is what Puttenham was describing in his diagrams, there are many more rhyme schemes. We can quickly see that the problems of five-line rhyme schemes and incense stick combinations in *Genji-ko* are exactly the same because we are looking at which things in the set (of five sticks, or of five lines) match up. Puttenham was way behind the Japanese, though, because he said that there were just seven possible rhyme schemes for a five-line stanza, "whereof some of them be harsher and unpleasaunter to the eare then other some be," while every *Genji-ko* player would have known that there are in fact fifty-two possibilities.

Because of *Genji-ko*, mathematicians in Japan became interested in counting the number of ways you can break up a set of objects (incense sticks or anything else) into different parts well before Western mathematicians considered the problem. This number of ways is nowadays called the Bell number of the set. The Bell numbers grow very quickly. The fourth Bell number is 15 (the number of quatrain rhyme schemes), the fifth is 52, the sixth is 203, but the tenth is already 115,975. Actually, I think I experienced the sixth Bell number in harrowing detail after recklessly agreeing to host a summer sleepover for our then eleven-year-old daughter Millie, when it seemed that all 203 possible ways for a group of six preteen girls to split off into mutually antagonistic cliques were attempted over a single night. The Japanese mathematician Yoshisuke

Matsunaga found an ingenious way to calculate Bell numbers for any size set way back in the mid-eighteenth century, giving, for instance, the eleventh Bell number as 678,570. I don't know why these numbers are named after the twentieth-century Scottish mathematician Eric Temple Bell, who wrote a paper about them only in 1934. He himself made it clear in the paper that he was not the first to work on them and that they had been rediscovered many times. It's another example of Stigler's law of eponymy, which states that no scientific discovery is named after its inventor (a law that holds also for Stigler's law of eponymy).

Rhyme schemes are among the defining characteristics of poetic forms—sonnets, villanelles, alexandrines, and so on. A villanelle, for example, is a nineteen-line poem consisting of five three-line stanzas with the *aba* rhyme scheme and a final *abaa* quatrain. There is additional structure: the first and third lines of the opening stanza repeat, alternately, as the final line of successive stanzas and as the last two lines of the quatrain. Probably the most famous villanelle is "Do not go gentle into that good night," Dylan Thomas's wonderful anthem to the human spirit. Sonnets, meanwhile, consist of fourteen lines. There are different traditional rhyme schemes in different languages, but Shakespeare and most other English-speaking writers have used three *abab* quatrains, followed by a rhyming couplet.

Shakespeare was a prolific poet—the 1609 edition of his collected sonnets contains 154 of them. But this is nothing compared to the French author Raymond Queneau's *Cent mille milliards de poèmes*, which uses the mathematics of randomness to fit 100 trillion sonnets into a single book. How is this possible? Let me explain. Everyone loves a sonnet, but my editor would kill me if I wanted to include 100 trillion of them in this book, so I decided to prolong my life by giving a smaller example to set the scene. To that end, I've deployed my amazing poetry skills to write some limericks for you instead.

Limericks are short, usually humorous poems consisting of five lines with the rhyme scheme *aabba*, popularized in England in the nineteenth century by the Victorian writer Edward Lear. Here's a typical example from his bestselling 1861 *Book of Nonsense*:

There was an Old Lady whose folly,
Induced her to sit on a holly;
Whereon by a thorn,
Her dress being torn,
She quickly became melancholy.

Lear is sometimes called the Father of Limericks, although he didn't use the term "limerick" himself (it's first recorded in 1898) or even invent them. However, with his much-loved books he certainly popularized the form, writing an impressive 212 limericks along the way. It's rather unclear how they ended up being named for an Irish county. One theory is that the name arose from a particularly popular example (not one of Lear's) that featured the line "Will (or won't) you come to Limerick?"

With the amazing power of randomness, I hereby scoff at 212 limericks and present a means to writing many more with a minimum of effort and artistic ability. Here are two not very good limericks (shown on the left and on the right, below) that I've invented to show you the method:

There once was a woman called Jane
There once was a person from Maine

Who constantly traveled by train
Who never went out in the rain

When going abroad
Damp days left her bored

She couldn't afford
                    Oh how she adored

A wonderful journey by plane
                    A week in the sunshine in Spain

From these two starting points, you can construct many more limericks. You do it by randomly picking lines from the two choices you have at each point. You can, for example, toss a coin to determine each line. If it's heads, you read the left-hand line; if tails, the line on the right. Brilliantly, there is a website, justflipacoin.com, that allows you to do this even without taking the trouble to find a physical coin. I tried it just now and got heads, tails, tails, heads, tails. So my new limerick reads:

There once was a woman called Jane
Who never went out in the rain
Damp days left her bored
She couldn't afford
A week in the sunshine in Spain

Since the poem has to "work" whichever option you pick for each line, if you want to try doing something like this, you need to understand the structure of the poem. As I noted already, the limerick has the rhyme structure *aabba*, so you need three *a* rhymes in each limerick. That means for two limericks you'll need six *a* rhymes. In this toy example, I chose "Jane," "train," "Maine," "rain," "plane," and "Spain." If you wanted a third limerick you could weave in words like "drain," "pain," "complain," "feign," "rein," and so on.

Our little poem set of two limericks has two choices for each of the five lines. There are two possible first lines. Each of these can be followed by two possible second lines. This means we have $2 \times 2 = 4$ possibilities for the first two lines. Each of these can in turn be followed by two options for line 3, giving $2 \times 2 \times 2 = 8$ possibilities for the first three lines. At each stage, the number of possible poems doubles. With our five lines to choose, we end up with a total of $2 \times 2 \times 2 \times 2 \times 2 = 32$ bona fide limericks. But if we wrote just one more limerick, we'd have three choices for each line, meaning a total of

$$3 \times 3 \times 3 \times 3 \times 3 = 243$$

limericks. Here's a third limerick for your delectation:

There once was a girl from Bahrain
Who viewed snow and hail with disdain
The cold she abhorred
She cheered when she scored
A trip to the African plain

Congratulations, you are now the proud owner of thirty-one more limericks than are contained in the entire oeuvre of Edward Lear. If you can add a fourth limerick to this set, then the total number will leap to $4 \times 4 \times 4 \times 4 \times 4$, which is 1,024, and since I wrote only 243 of these, you are morally entitled to more than 75 percent of the worldwide fame that will surely result from the composition of over a thousand limericks.

We can now see just how Raymond Queneau managed to construct his 100 trillion poems. It's exactly the same principle, just on a bigger scale. The poems are sonnets, so they have 14 lines. Queneau chose the rhyme scheme *abab abab ccd eed.* (Translations into English have tended to use the Shakespearean *abab cdcd efef gg.*) *Cent mille milliards de poèmes* consists of ten sonnets, printed on ten consecutive sheets. All the first lines rhyme with each other, all the second lines rhyme with each other, and so on. In effect, the ten sonnets line up to create a three-dimensional poem. This means, for instance, that of the 140 total lines, 40 of them, 4 in each poem, must end with rhyme *a*. Sonnets can then be made by choosing any of 10 possible lines at each point. So I might choose line 1 from poem 3, line 2 from poem 1, line 3 from poem 4, and so on. If I continued selecting the poem numbers by following the digits of π, nobody could stop me from then saying I have produced a πem (sorry).

How many poems are contained in this little book, then? Well, the number of possible first lines is 10. Each can be followed by any one of ten second lines, giving $10 \times 10 = 100$ possibilities for the first two lines. With fourteen lines altogether, the total number of possibilities is ten multiplied by itself fourteen times, or 100,000,000,000,000. In other words, 100 trillion. Is this the longest book ever written? If you read a different sonnet every minute without stopping, it would take 190,128,527 years to read them all. (Raymond Queneau did this calculation too, but he arrived at an answer of 190,258,751 years, which made me doubt my arithmetic skills. But a quick check shows that his is the answer you get if you read one sonnet per minute but forget about leap years. Perhaps Queneau was very generously allowing his readers to take a day of rest on February 29.) A philosopher might ask: Did Queneau write all these poems? In what sense do they exist at all? I don't know, but Queneau was a member of a group of writers and poets experimenting with what they termed "potential literature." This group was known as the Oulipo—I'll be showing you more of their work and ideas later. But a book of 100 trillion poems is certainly an excellent example of potential literature.

The mathematics of poetry does not stop with rhyme schemes; wherever there is structure, there is mathematics, and rhyme schemes are just one way to impose structure. If we abandon rhyme, then something else needs to take its place. One possibility that dates back to medieval times is the sestina, and I want to talk about this form in particular because its elegant structure works thanks only to some curious mathematics involving the number six.

A sestina consists of six stanzas, each of six lines. The last words of each line in the first stanza reappear as the last words of the lines in subsequent stanzas, in a different (but specific) order. Then the whole thing is usually finished with a three-line "envoy" that features all six end-words somewhere in it.

I'd like to give you a complete example, if I may, so that you can see what is going on. There's a lot of choice, because even though this form was first used more than eight hundred years ago, it is still in use and has enjoyed periods of great popularity. The 1950s were even described as the "age of the Sestina" by James Breslin (at the time, a professor of English at UC Berkeley). There are sestinas by poets from Dante to Kipling, Elizabeth Bishop to Ezra Pound, through to contemporary works by the American poet David Ferry ("The Guest Ellen at the Supper for Street People") and by the English "thingwright"—this is the marvelous way she describes herself on her website—Kona Macphee (the desperately sad 2002 poem "IVF"). The example I've chosen is a poem by Charlotte Perkins Gilman, who is best known nowadays for her 1892 short story, "The Yellow Wallpaper."

To the Indifferent Women
A Sestina
by Charlotte Perkins Gilman

You who are happy in a thousand homes,
Or overworked therein, to a dumb peace;
Whose souls are wholly centered in the life
Of that small group you personally love—
Who told you that you need not know or care
About the sin and sorrow of the world?

Do you believe the sorrow of the world
Does not concern you in your little homes?

That you are licensed to avoid the care
And toil for human progress, human peace,
And the enlargement of our power of love
Until it covers every field of life?

The one first duty of all human life
Is to promote the progress of the world
In righteousness, in wisdom, truth and love;
And you ignore it, hidden in your homes,
Content to keep them in uncertain peace,
Content to leave all else without your care.

Yet you are mothers! And a mother's care
Is the first step towards friendly human life,
Life where all nations in untroubled peace
Unite to raise the standard of the world
And make the happiness we seek in homes
Spread everywhere in strong and fruitful love.

You are content to keep that mighty love
In its first steps forever; the crude care
Of animals for mate and young and homes,
Instead of pouring it abroad in life,
Its mighty current feeding all the world
Till every human child shall grow in peace.

You cannot keep your small domestic peace,
Your little pool of undeveloped love,
While the neglected, starved, unmothered world
Struggles and fights for lack of mother's care,
And its tempestuous, bitter, broken life
Beats in upon you in your selfish homes.

We all may have our homes in joy and peace
When woman's life, in its rich power of love
Is joined with man's to care for all the world.

Let me show you how a sestina is constructed. To move from one stanza to the next, you move around the end-words in precisely the same way each time, a sort of ordered disorder created by working in reverse from the last end-word backward, and interleaving them with the first end-words in the right order, until we've used them all up. We can see this in Charlotte Perkins Gilman's sestina. The end-words in the first verse are homes/peace/life/love/care/world. Reversing the last words gives world/care/love . . . , and we interleave these with homes/peace/life . . . , so as to obtain

world care love
homes peace life

That is, world/homes/care/peace/love/life. And these are, as you can see, exactly the end-words in the second stanza. This specific shuffling gives a nice continuity between the stanzas, because the end of the last line in one stanza is the end of the first line of the next. The structure continues, though, because we repeat this same reverse interleaving on the end-words of the second stanza to obtain the ordering of the end-words in the third stanza. If you try this, you'll find that it turns world/homes/care/peace/love/life into life/world/love/homes/peace/care. And we repeat this process to obtain the orderings for the fourth, fifth, and sixth stanzas. There is a beautiful bit of unseen structure here, too, in that if we were to continue to a seventh stanza, our interleaving process applied to the sixth stanza's ordering of peace/love/world/care/life/homes would result in the end-words homes/peace/life/love/care/world. If this looks familiar, it should—it's the same ordering as we started with. The six stanzas therefore give us, even though we don't consciously recognize it, a complete circle of six iterations, which if continued would bring us exactly back to our starting point. I think we do experience and appreciate this mathematical structure subconsciously, even though we may not detect it consciously. The shuffling also has pleasing internal symmetries—every end-word appears at the end of every different possible line, from first to last, in precisely one stanza. It's a compelling design.

Unusually for so ancient a form, we have a plausible candidate for who invented it—the twelfth-century poet Arnaut Daniel. It was viewed as a very refined form of poetry that only the expert troubadour could master. I don't know how Daniel came up with the idea—it's a really simple permutation, very easy to remember, and you might think, once you hit on the process to follow, that given that the number of stanzas and the number of lines in each stanza are equal, both six, then you'll naturally come back to where you started with after six shuffles. But let's see what happens when we try to create a "quartina" with the same process. We start with a four-line stanza. Let's suppose our end-words are north/east/south/west. Remember the rule—we work in reverse order from the end, interleaving with words from the start. So we get west/north/south/east for our second stanza. We repeat the process to get east/west/south/north for the third stanza, then again to get north/east/south/west for the fourth stanza. Oh, no! We have regained our original order in the fourth stanza! So this process would not give us four different stanzas. Even worse, you can see that the end-word "south" gets stuck—it's the end-word of the third line in every stanza.

If you try to create a sestina-like poem with numbers other than six, you'll find that sometimes it works and sometimes it doesn't. In the 1960s, people started to try to figure out which values of n work. These "generalized sestinas" were named *queninas* by the Oulipo, in honor of Raymond Queneau. It turns out to be a really tricky problem. It works, for instance, for 3, 5, 6, 9, and 11, but not for 4, 7, 8, and 10. Amazingly, it is still an unsolved problem whether there are infinitely many values of n for which a quenina is possible, although a 2008 paper by the mathematician Jean-Guillaume Dumas described exactly the properties that such n would have to have. There is a particularly nice kind of number that will always have a quenina, a prime number called a Sophie Germain prime. It was named after a remarkable mathematician who did brilliantly innovative work in several areas of mathematics despite having to register at university under a false name and get other students to send her the course notes, due to the dreadful failing of being a woman—this was eighteenth-century Paris, after all. A prime number is called a Germain prime if, when you double it and add 1, the answer is again prime. The number 3, for example, is a Germain prime because $2 \times 3 + 1 = 7$ is again prime, but 7 is not a Germain prime because $7 \times 2 + 1 = 15$ is not prime. I can't prove it for you, but it turns out that a quenina is possible for every Sophie Germain prime, which I love. Indeed, I know of at least one published "tritina" (three stanzas of three verses; the envoy is one line that includes all three of the end-words), by the English poet Kirsten Irving.

Talula-Does-the-Hula-from-Hawaii
by Kirsten Irving

Where do stupid names end up, these shorn tags
tied on toes by parents with the abandon
and foresight of tyrants annoying their court?

Today the three of you, now strangers, leave court
in opposite directions, untying cloakroom tags
from belongings, as you abandon

what passed for a name. That punchline abandoned
to the playground's corrupt court
and the toilet wall's smeared tags.

Tags abandoned, a girl who's not Talula courts the world.

Rhyme schemes and queninas impose structure on the ending of lines, and they already give us some fascinating mathematics to play with. But there's even more to explore when we consider the patterns *within* lines of poetry, and that's what we'll turn to next.

In addition to the rhyme scheme, poetic forms often have a specific rhythm in their lines, which we call meter. Shakespeare's plays are full of iambic pentameters, for instance. The "penta" bit is from the Greek word for five, and an "iamb" is a two-syllable phrase of which the second syllable is stressed. Thus an iambic pentameter has ten syllables, with the second one in each pair being stressed. I've underlined the stressed syllables in the following example, from the balcony scene in *Romeo and Juliet.*

> But soft, what light through yonder window breaks?
> It is the East, and Juliet is the sun.

This "di-dum di-dum di-dum di-dum di-dum" can be represented visually using dots and dashes, just like Morse code. An iamb is ·–, and an iambic pentameter looks like this:

·–·–·–·–·–

The basic patterns of stressed and unstressed syllables are called *feet.* Two common examples, along with the iambs we have just seen, are trochee (–·), as in "Quoth the Raven 'Nevermore,'" and dactyl (–··), as in "The Lost Leader," by Robert Browning, which begins, "Just for a handful of silver he left us"—actually this is three dactyls and a trochee at the end. How many possible meters are there for a given number of syllables? There are two possibilities for each syllable—stressed or unstressed—so the number of one-syllable feet is two (· or –). To get to two syllables, we can add either a · or a – to either of these, so the total is four. We can add a · or a – to each of these four to get eight possible three-syllable meters, and it just keeps doubling—we end up with a sequence 1, 2, 4, 8, 16, and so on, the powers of 2.

But there's a form of poetry in which something very different happens. I first read about it in Jordan Ellenberg's excellent paean to geometry, *Shape*. He recounts how a mathematician friend, Manjul Bhargava, told him about the meters of Sanskrit poetry. As in English poetry, the pattern of syllables is important, but while with English we look at where the stresses lie, in Sanskrit it's the *length* that matters. Syllables are either *laghu* (light) or *guru* (heavy). Crucially, *laghu* syllables count as one unit, and *guru* as two. This means it's a bit more complicated to work out, for instance, how many four-syllable meters are possible. We can't just take the number of three-syllable meters and double it. So what do we do? Well, there's just one one-syllable possibility: *laghu*. There are two two-syllable options: *laghu laghu*, or *guru*. For three syllables, you can check that the three possibilities are *laghu laghu laghu*, *laghu guru*, or *guru laghu*. For four syllables, let's get a bit clever and divide the problem into two. Either the meter starts with *laghu*, or it starts with *guru*. If it starts with *laghu*, then we can choose from any of the three three-syllable meters to add on to it, to arrive at four syllables. If it starts with *guru*, then we can choose either of the two two-syllable meters to add on. So the total is 3 + 2 = 5:

laghu laghu laghu laghu
laghu laghu guru
laghu guru laghu
guru laghu laghu
guru guru

What's more, you can always play this trick. Five-syllable meters are either *laghu* + (a four-syllable meter) or *guru* + (a three-syllable meter). So the number of five-syllable meters equals the number of four-syllable meters plus the number of three-syllable meters, which is $5+3=8$. We can carry on like this. The next number is just the sum of the previous two numbers. So we get a Sanskrit meter sequence like this:

$$1, 2, 3, 5, 8, 13, 21, \ldots$$

You may have encountered this sequence before. It's better known in English-speaking countries as the Fibonacci sequence, popularized in Europe in the thirteenth century by Leonardo of Pisa, whose nickname was Fibonacci. (Sometimes it's shown as beginning with two 1s, but it's the same basic principle.) Each term after the first two, as we've said, is the sum of the two previous terms. For example, $13=5+8$. The next term in the sequence after 21 will therefore be $13+21=34$. The Fibonacci sequence has many interesting properties. One is that the sequence $\frac{2}{1}, \frac{3}{2}, \frac{5}{3}, \frac{8}{5}, \frac{13}{8}, \frac{21}{13}, \ldots$ of ratios of consecutive terms converges to the famous *golden ratio* $\frac{1+\sqrt{5}}{2} \approx 1.618$.

When Fibonacci introduced the sequence in his 1202 book *Liber Abaci* ("The Book of Calculation"), it was in the context of a rather fatuous puzzle about rabbits. You start with one breeding pair of newborn rabbits. A breeding pair mates after one month, and the female gives birth to a new breeding pair one month after that. Rather unrealistically, the rabbits never die, they keep breeding forever, and we have to ignore minor concerns like rabbit incest. The question is, how many pairs of rabbits are there after one year? We can see that the same rule applies to this sequence. In any given month, the total number of pairs will be the number there was a month ago plus the number of newborn pairs, which (since it takes two months from birth to produce a new pair) is the number of pairs there were two months ago. So each term is the sum of the previous two terms. But this sequence had been known to poetry scholars in India for centuries before Fibonacci. The metrical experts Virahanka (sometime between 600 and 800 CE), Gopala (sometime before 1135 CE), and Hemachandra (around 1150 CE) all knew the sequence and how to produce it, and there's some evidence that it was known even earlier, in the writings of Pingala (around 300 BCE). Perhaps it's time to rename the Fibonacci numbers.

Mathematics and poetry are two of our most ancient forms of creative expression, and their connections reach back to the very beginnings of writing itself. The earliest known works by a named author in the whole of human history were created by a remarkable woman named Enheduanna, who lived over four thousand years ago in the Mesopotamian city of Ur. She wrote perhaps the very first collection of poems—a cycle of forty-two "Temple Hymns." But as high priestess of the moon god Nanna, she would have needed knowledge of astronomy and mathematics as well. These come together in her poetry, both in her use of numbers, particularly the number seven, and in mention of calculation and geometry. The final Temple Hymn speaks of the mathematical activities of the "true woman of unsurpassed wisdom":

She measures the heavens above
and stretches the measuring cord on the earth.[6]

From these earliest beginnings, the love affair between poetry and mathematics has flourished. Mathematics has been there in the deep currents of verse, underpinning its rhymes and hidden in its structures. As the great nineteenth-century mathematician Karl Weierstrass wrote, "A mathematician who is not somewhat of a poet, will never be a perfect mathematician." And poetry? It's simply the continuation of mathematics by other means.

2

The Geometry of Narrative

How Mathematics Can Structure a Story

At a public lecture in 2004, Kurt Vonnegut gave illustrations of the "graphs" of some possible stories.[1] The first of these was "Man in a Hole":

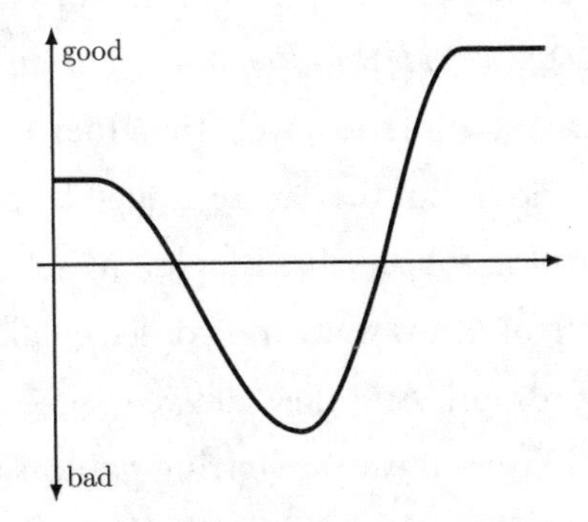

In Vonnegut's graphs, the vertical axis measures good fortune and the horizontal axis measures time passing—a rising curve means improving fortunes, a falling curve means things are getting worse. In "Man in a Hole," for instance, we start with someone going along happily when suddenly disaster strikes, but everything works out wonderfully in the end. A novel in this category might be *David Copperfield*—or, to give it its full, glorious title, *The Personal History, Adventures, Experience and Observation of David Copperfield the Younger*

of Blunderstone Rookery (Which He Never Meant to Publish on Any Account). The young David has a very happy childhood until he is seven, when his mother first marries beastly Mr. Murdstone, then soon afterward dies, leaving poor David orphaned. But after many reverses and trials, David eventually finds happiness. Vonnegut gave three other graphs, which I've sketched below:

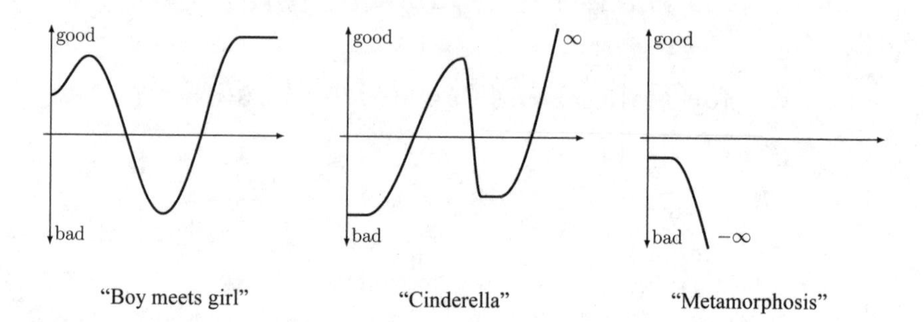

"Boy meets girl" "Cinderella" "Metamorphosis"

"Boy meets girl," of course, is a feature of most romantic novels. Boy meets girl, boy loses girl, boy gets girl in the end. Happiness all around. To pick a random example, take the story line of Jane Bennet and Mr. Bingley in Jane Austen's *Pride and Prejudice*. Jane and Bingley are fairly content already at the start of the novel. Then they meet and fall in love, and life looks even better. But they are separated by the machinations of proud Mr. Darcy and snobbish Miss Bingley. Misery ensues. In the end Darcy realizes the error of his ways and confesses all to Bingley, who at once returns to get his girl. And they all live happily ever after.

In "Cinderella," by contrast, the starting point is unhappiness. Poor Cinders sleeps in the ashes of the fire (hence her name) and works all day for her horrid stepsisters. But then things start looking up. Off she goes to the ball, where she meets Prince Charming, but then—disaster! Midnight strikes and all appears lost. Fortunately, her feet are so freakishly shaped that she's the only girl in the kingdom who can fit into the glass slipper left behind when she fled. She marries the prince, and her happiness becomes infinite.

The last of Vonnegut's graphs is "Metamorphosis," which refers to the darkly comic story by Franz Kafka. You will remember that this is the tale of Gregor Samsa, unhappy and alienated in his job as a traveling salesman. One morning he wakes up and finds that during the night he has turned into a gigantic "vermin" (usually assumed to be a cockroach). There follows a degrading and painful descent into illness and death. Good old Kafka.

We might place works like *The Metamorphosis* at the pessimistic end of the fine tradition of absurdism in literature, a style of writing amusingly described by author Patricia Lockwood as "novels where a man turns into a teaspoonful of blackberry jam at a country house."[2] For a truly absurd story graph, there's no better place to turn than the brilliant, anarchic work of genius that is *Tristram Shandy*. Laurence Sterne's novel appeared originally in nine volumes, published over eight years from 1759 to 1767. The narrator is Tristram Shandy, a gentleman who has decided to write his autobiography but is continually thwarted in this aim by the intrusions of other characters into the story. Tristram gets sidetracked by so many digressions and diversions that he doesn't even manage to be born until Volume III. It's a joyously chaotic read. Toward the end of Volume VI, Tristram Shandy draws a diagram of his narrative "lines" so far:

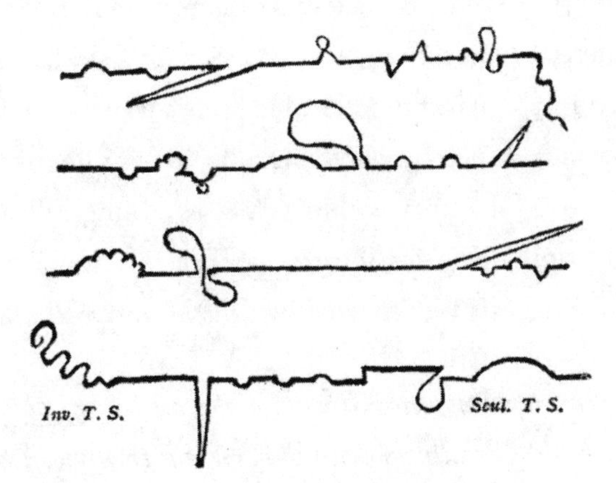

"These were the four lines I moved in," he writes, "through my first, second, third, and fourth volumes. In the fifth volume I have been very good,——the precise line I have described in it being this:"

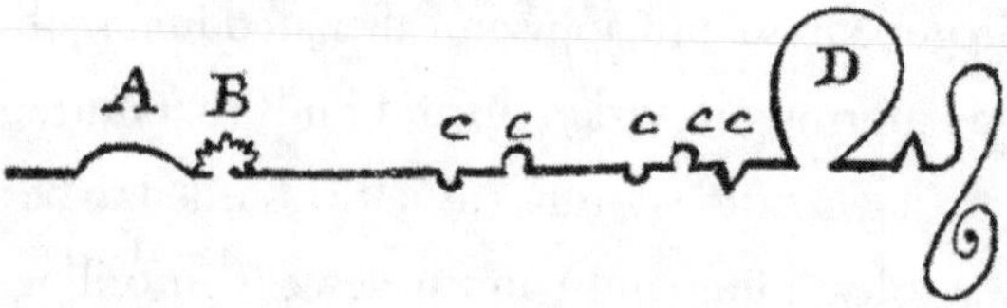

He claims that this is an improvement: "except at the curve, marked A, where I took a trip to Navarre,—and the indented curve B, which is the short airing when I was there with the Lady Baussiere and her page,—I have not taken the least frisk of a digression, till John de la Casse's devils led me the round you see marked D.—for as for c c c c c they are nothing but parentheses." "If I mend at this rate," he says, "it is not impossible but I may arrive hereafter at the excellency of going on even thus:

which is a line drawn as straight as I could draw it. . . . The best line! say cabbage planters—is the shortest line, says Archimedes, which can be drawn from one given point to another." You will be pleased to hear that this optimistic prediction proves entirely false, and the last volumes of the novel romp around as gleefully as the first.

Vonnegut's graphs and Shandy's crazy narrative "lines" are amusing, but are there more sophisticated, genuinely mathematical takes on narrative and plot? This chapter takes its title from Hilbert Schenck's story "The Geometry of Narrative" (1983), in which a student suggests that simple plot "lines" are just the start. He finds a way to link Shakespeare's *Hamlet* to a four-dimensional "hypercube" by arguing that we should think of instances of a story within a story as adding a dimension. That is, instead of time being the fourth dimension, Schenck's protagonist, Frank Pilson, suggests we use what he calls *narrative distance*:

> *Here are two separate three-dimensional realities: the play,* Hamlet *itself, with old Claudius popping his mental cork when he sees the Hamlet-buggered script acted out, and the shorter, smaller, on-stage murder-of-Gonzago play. But the little play is at a greater distance from* Hamlet, *both from the real audience and from the Court of Denmark watching it on stage, since it is presented as a created artefact*

within the "true" or "real" drama. So not only is this part of Hamlet *modelled by a four-dimensional geometrical object, but the staging assumes the exact projected form of the hypercube, with one small stage located in the middle of the other, larger one.*

The rest of the story very cleverly sees the narrative camera repeatedly zooming in, so to speak, so that the frame of reference is constantly shifting. The part of the story you encounter first can change your understanding of the plot—is the story a first-person account by Pilson, telling us about his literature seminar and quoting excerpts from a story, or are we actually reading a story about an author who happens to be working on his novel about a student called Pilson? Our understanding of these different levels of narrative may prompt us to revisit the text and read it again, but from a different viewpoint or in a different order.

Shakespeare was not thinking of hypercubes when he wrote *Hamlet*, but many authors have consciously chosen to impose mathematical constraints on their narratives. As the author Amor Towles said in a 2021 interview,[3] "Structure can be very valuable in artistic creation. Much as the rules of the sonnet are valuable to the poet, adopting the rules and trying to invent, within those rules, something that's new and different, the structure of a novel can do the same kind of thing." You may be thinking, *Why would a writer bother with some fancy structure? Why not just write a good story?* This, I would argue, is a false dichotomy. All writing has structure from the get-go. Language itself is built of component parts, each of which has patterns. Letters make up words, words form sentences, sentences form paragraphs, and so on. This is already a structure, analogous to the hierarchy of point, line, plane in geometry. At each stage, further structures can be imposed. Paragraphs, for instance, can be joined together to form chapters. The decision is not *whether* to structure your work; rather it's what structure to choose. Within each of these levels, writers may choose to add additional structural constraints. This added structure works best when it feels most natural, when it fits with the narrative themes or the design of the plot.

Let's start at the highest level usually used in novels: the chapter. Eleanor Catton's *The Luminaries*, which was published in 2013, is an astonishing achievement. Catton was the youngest finalist ever for the Booker Prize. She then became, at twenty-eight, its youngest winner ever. The judges described the book as a "dazzling," "luminous" work, "vast without being sprawling." And it is indeed vast. At 832 pages, it was the longest book ever to win the prize. The events of the novel are centered on the gold rush town of Hokitika, New Zealand, in the mid-1860s. The first chapter, mathematically titled "A Sphere Within a Sphere," opens with the prospector Walter Moody arriving in Hokitika on January 27, 1866, and walking in on a meeting of twelve local men who have gathered to discuss a series of recent crimes. He becomes entangled in a web of murder, strange disappearances, attempted suicide, opium dealing, and the discovery of £4,096 worth of stolen gold.

There are twelve chapters, or parts, each taking place over the course of a single day in 1865 or 1866 (the novel's first chapter begins, chronologically, at the midpoint of events). The twelve men whom we meet at the start of the novel are each associated with a specific sign of the zodiac. Their actions and behavior in each of the twelve chapters are determined in part by that sign's astronomical configuration on the date of the chapter. Catton did careful research into the positions of the stars and planets in the night sky of Hokitika on those precise dates. By the way, I don't think that this is because she is necessarily a believer in astrology. She says of Walter Moody that he was "not superstitious, though he derived great enjoyment from the superstitions of others." The astrological and astronomical information is both a way to give structure and a way to inform the broader meditation in the book about the interplay between fate, circumstance, and free will.

In *The Luminaries*, each chapter is divided into a specific number of sections, and in every case, the number of sections, added to the number of the chapter, is the same: thirteen. Thus the first chapter has twelve sections, the second chapter has eleven sections, and by the time we get to the twelfth and final chapter, it has just one section. This kind of pattern, in which we see the same increase or decrease each time, as in the sequence 12, 11, 10, 9, . . ., is known in mathematics as an *arithmetic progression*. Hidden in the thirteenness of chapter number plus number of sections is a really simple trick to add up the total number of sections in the book. It would be annoying to have to calculate the sum $1+2+\cdots+12$ by laboriously adding the numbers one by one. But if you go over all twelve chapters, in each case we know that the chapter number plus the number of sections equals 13. So the total of these thirteens, over the twelve chapters, is $12\times13=156$. This picture shows, on the left, the chapter numbers, and on the right, the sections, adding up to 13 every time.

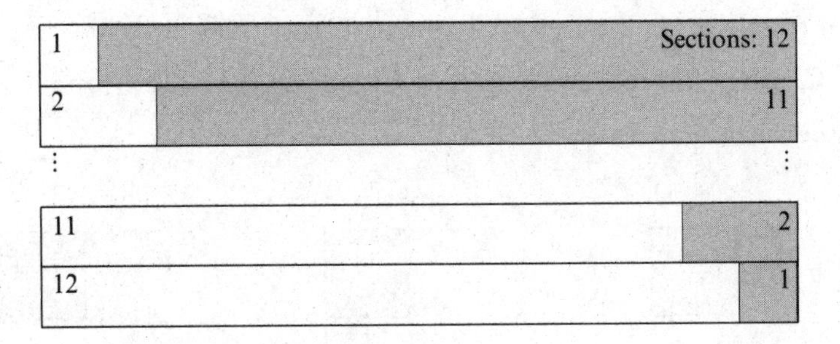

But this total is double what we need, because it's also got the $1+2+\cdots+12$ of the chapter numbers in there. All we have to do is halve it: the total number of parts is $\frac{1}{2}(12\times13)=78$.

This trick is one of my first mathematical memories—my mother taught it to me when I was a kid, and I thought it was pretty amazing. She recounted the (possibly apocryphal) story of how the great mathematician Carl Friedrich Gauss, while still in elementary school, ruined a teacher's attempt to get a bit of peace and quiet one afternoon, when the teacher set Gauss's class the task of adding up all the numbers from 1 to 100. The young Carl apparently invented on the spot this little trick I've just explained. If our book had 100 chapters with the same pattern, then the sum $1+2+\cdots+100$ is $\frac{1}{2}(100\times101)=5{,}050$. Cool! I feel a bit sorry for the poor teacher, though—all he wanted was half an hour of quiet.

The most interesting and impressive aspect of the mathematical structure of *The Luminaries* is that each chapter is half the length of the last. That constraint has significant implications for the length of the novel. We can represent the length of the first chapter with a rectangle (we might measure the length in words, characters, lines, pages—whatever you prefer; it doesn't make much difference). Here it is:

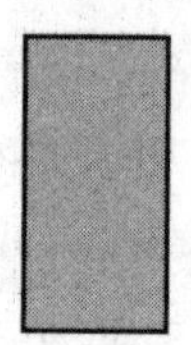

Now, the next chapter is half as long, so we can slot it in with a half-sized rectangle on the right. Chapter 3 is half the length of Chapter 2, and Chapter 4 is half the length of Chapter 3. I've shown the first few chapters in the picture:

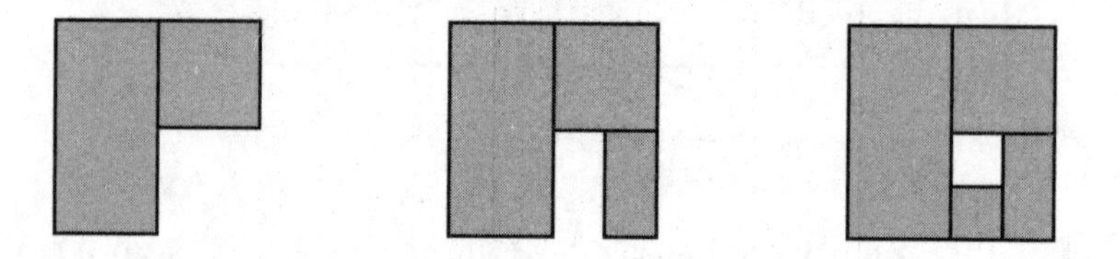

We can keep on going, slotting ever smaller and smaller rectangles into this picture and never escaping from the outer square boundary. I've added Chapters 5, 6, 7, and 8 to the left-hand diagram and Chapters 9 to 12 to the right-hand diagram just to show you.

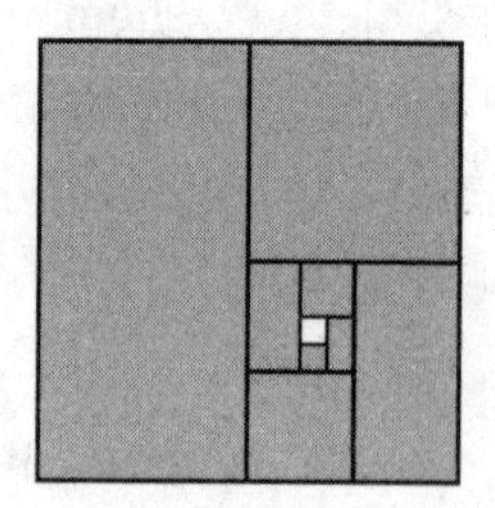

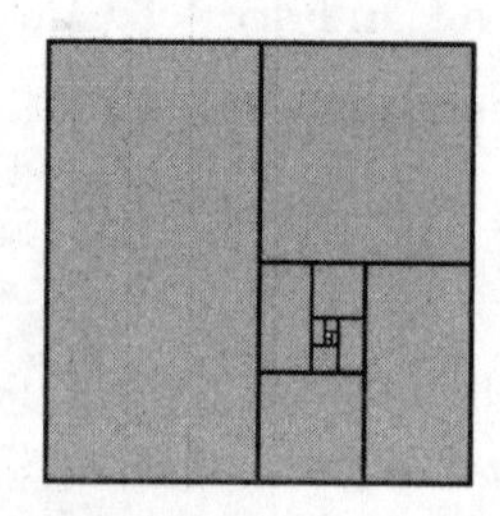

We are creating a beautiful spiral effect, with each consecutive chapter fitting nicely into the ever smaller space remaining. What this means is that however many chapters there are, the total length of the book is less than twice the length of the very first chapter! There's no escape—even if you had a million chapters.[4]

We know that this book has twelve chapters. Is there some nice, easy trick, the way there was for counting sections, to tell us exactly how long the book will be once we know how long the first chapter is? Happily, there is. The kind of sequence we see here with chapter lengths, namely $1, \frac{1}{2}, \frac{1}{4}, \frac{1}{8}, \cdots$, is one in which to get from one step to the next, we don't add or subtract a fixed amount; we rather *multiply* by a fixed amount (in this case $\frac{1}{2}$). It's known as a *geometric progression*, and the trick for adding its terms involves an ingenious idea. I'll show you for the case in which we halve successive terms, because that's what we are doing with chapter length, but the same kind of idea works much more generally.

Okay, so let's say our first chapter has length L, where L is measured however you like: pages, words, whatever. Then Chapter 2 has length $\frac{1}{2}L$. Chapter 3 has length $\frac{1}{4}L$, and so on. The total length of the book is going to be

$$L+\frac{1}{2}L+\frac{1}{4}L+\frac{1}{8}L+\frac{1}{16}L+\frac{1}{32}L+\frac{1}{64}L+\frac{1}{128}L$$
$$+\frac{1}{256}L+\frac{1}{512}L+\frac{1}{1024}L+\frac{1}{2048}L,$$

which we can make a bit simpler by bringing out that factor of L to get

Book Length

$$=L\left(1+\frac{1}{2}+\frac{1}{4}+\frac{1}{8}+\frac{1}{16}+\frac{1}{64}+\frac{1}{128}+\frac{1}{256}+\frac{1}{512}+\frac{1}{1024}+\frac{1}{2048}\right).$$

Here's the trick. Halve both sides:

$$\frac{1}{2}(\text{Book Length}) = L\left(\frac{1}{2}+\frac{1}{4}+\frac{1}{8}+\frac{1}{16}+\frac{1}{64}+\frac{1}{128}+\frac{1}{256}+\frac{1}{512}+\frac{1}{1024}+\frac{1}{2048}+\frac{1}{4096}\right).$$

See how there's a $\frac{1}{2}$ in each expression, lining up with each other, and a $\frac{1}{4}$, and so on all the way to $\frac{1}{2048}$? I'm now going to subtract the second expression from the first. So on the left-hand side we'll have the total book length, take away half the book length, leaving half the book length. On the right, almost everything will cancel out and we'll just get

$$\frac{1}{2}(\text{Book Length}) = L\left(1-\frac{1}{4096}\right).$$

Doubling up, here is our patented formula for finding the length of *The Luminaries*: it's $2L\left(1-\frac{1}{4096}\right)$. Remember that £4,096 of stolen gold? There it is—sewn into the very fabric of the book!

The choice to have twelve chapters fits in very nicely with the other structural elements of the book, but as I'm about to show you, the number of possible chapters is very closely constrained by our geometric progression. Looking a bit more carefully at our chapter lengths, they are related to powers of 2. The mathematical notation for powers is to put a little superscript after the number to mean the number of times it's multiplied by itself. For instance, 2^5 means $2 \times 2 \times 2 \times 2 \times 2$, which is 32. To get the length of, say, the seventh chapter of our book, we have had to halve our first chapter six times. This means the length of the seventh chapter is $\frac{1}{2^6}L = \frac{1}{64}L$, as you can see if you look back at my equation for the total length of the book. The twelfth and final chapter has length $\frac{1}{2048}L = \frac{1}{2^{11}}L$. Let's say that the shortest chapter has length S. Then $L = 2^{11}S = 2048S$ in the case of the twelve-chapter *Luminaries*. The total length for our twelve-chapter book was $2L\left(1-\frac{1}{4096}\right)$. Replacing L with

$2^{11}S$ we get $2\times 2^{11}S\left(1-\frac{1}{4096}\right)$, and noticing that 2×2^{11} is just 2^{12}, which is 4,096, this all simplifies beautifully to $(2^{12}-1)S$.

Putting actual numbers in, it just takes a moment to calculate that $2^{12}-1=4096-1=4095$. What this means is that the total length of the book is 4,095 times the length of the final chapter. It's pretty obvious, then, that the lengths can't be measured in pages, or even if the last chapter was just one page long, poor Eleanor Catton would have had to write a 4,095-page behemoth. *The Luminaries* is long, but not that long. Come to think of it, this explains why the recent TV adaptation did not follow the book's structure with twelve episodes, each half the length of the previous one: if the final episode were just one minute long, the first episode would have had to last over thirty-four hours.

It's quite hard to bind a book of much over a thousand pages, and probably even harder to find a publisher willing to print it, so let's take a thousand pages as a reasonable maximum, with about 400 words of text on each page. Then a sensible upper limit to work to would be 400,000 words. Even if the shortest chapter has just 100 words, then the total word count would be $100\times 4{,}095$, which is 409,500—that pushes the limits of what is attainable. I have just counted the words in the final twelfth chapter of *The Luminaries*, and there are 95. This gives an estimated 389,025 words. I do not claim this as an exact figure. There is some wiggle room because there are different ways of counting the words (do I count chapter headings? Do I count the words "Part Twelve" and so on?). With a shortest chapter of 95 words, there's no way the book could have more than twelve chapters—if it had thirteen chapters, for example, the word count would more than double to 778,145, which would definitely raise eyebrows at the printing press!

If someone really wanted to write a book with the chapter-halving property that had more than twelve chapters, what's the most chapters they could write? To find the total length of a book with (let's say) n chapters, in terms of the length of the shortest chapter, we can repeat the exact same calculation as we did for twelve chapters. If we had n chapters, the final chapter would have length $S = \frac{1}{2^{n-1}}L$, or equivalently $L = 2^{n-1}S$. The total length of the book would be not be $(2^{12} - 1)S$ but $(2^n - 1)S$. Even if the shortest chapter was just one word long, then an upper limit is reached pretty fast. To stay within 400,000 words, we could find the maximum number of chapters by solving $2^n - 1 \leq 400{,}000$. If you do this, you find that the highest n can be is 18. The last six chapters aren't really worth having—they would contain a total of just 63 words among them.

Why did Catton use this particular structure? Part of what makes it, and the novel, successful is that it is not a random choice. If you wanted something to do with twelve in your book, to emphasize the link to the twelve signs of the zodiac, you could make each sentence twelve words long, or have twelve chapters and $12^2 = 144$ sections, or all sorts of other possibilities. The decision to halve these twelve chapters each time, like the waning of the moon, works because it echoes both the astronomical and astrological themes of the book, as well as the development of the plot and the underlying central story of the two lovers, represented by the sun and moon. There are resonances in the text—things doubling and halving, falling and rising, increasing and decreasing like the sun, moon, stars, and the fates of the characters. The prostitute Anna Wetherell, despairing over the fact that her debts have doubled in the last month, reflects that "a woman fallen has no future; a man risen has no past."

We feel the tension rising as the parts become more condensed. In a 2014 interview, Catton said, "I see it like a wheel, a huge cartwheel, creaky at the beginning and spinning faster and faster as it goes." The sense of the inescapability of our fates increases as the constraints become ever tighter with each chapter—we saw the literal spiraling-inward effect—drawing us into the final, tender scene between the doomed lovers in the last chapter, Part Twelve. It is titled "The Old Moon in the Young Moon's Arms," and it takes place on January 14, 1866, just days before the events of Part One. This is the true center of the novel, and the spiraling progression we have seen calls to mind Yeats's image of the "widening gyre" from his unforgettable poem "The Second Coming," the first four lines of which read:

> Turning and turning in the widening gyre
> The falcon cannot hear the falconer;
> Things fall apart; the centre cannot hold;
> Mere anarchy is loosed upon the world

In the poem we follow the path of the gyre as it travels outward from the center in the whirling storm. But in *The Luminaries* we are following that path in reverse, closing further and further in upon the center. Appropriately enough then, for a novel with so many astrological references, *The Luminaries* shows us the widening gyre not straight on, but in retrograde.

In *The Luminaries*, the geometric progression structure is manifested in the physical length of chapters. But there's another kind of structure in every narrative: not spatial but chronological. In a novel, as E. M. Forster said, there is always a clock. Sometimes the ticking is very loud. Aleksandr Solzhenitsyn's *One Day in the Life of Ivan Denisovich* is precisely that—it recounts events during a single day of a ten-year sentence in the Russian gulag. Virginia Woolf's *Mrs. Dalloway* and James Joyce's *Ulysses* also take place over a single day, which just goes to show that the imposition of a constraint does not have to restrict creativity—three more different books are difficult to imagine. A still shorter time period is the basis of a poignant 2019 novel by the Turkish writer Elif Shafak. A woman named Leila is brutally murdered, and as her brain starts to shut down, memories from her life pass through her mind until her soul ultimately departs from her body. The length of time that passes during these strange liminal moments gives the book its title: *10 Minutes 38 Seconds in This Strange World.* If you are thinking that at this rate I'm going to claim there's a book in which no time passes at all, you'd be right. *Life: A User's Manual*, by the French author Georges Perec, purports to take place at a single moment: just after 8 p.m. on June 23, 1975.

The 2016 novel *A Gentleman in Moscow*, by Amor Towles, goes to the other extreme. Instead of a single day, its events take place over the course of thirty-two years. But it has a very sophisticated chronological framework. It's perhaps not surprising to find mathematical structures in the work of a writer who spent twenty years working as a Wall Street banker before his first novel (2011's bestselling *Rules of Civility*) was published. The best fact I've ever heard about Towles is that when he was ten years old he put a message in a bottle and threw it into the sea at a place called West Chop, Massachusetts. "If this gets to China," he wrote (or words to that effect), "please write back." How many children have done this, and how few have ever had a response? But a few weeks later, someone did write back, albeit not from China. Harrison Salisbury, then the managing editor of *The New York Times*, no less, had found the bottle, and the pair corresponded for several years, finally meeting when Towles was eighteen. Salisbury actually makes a cameo appearance in *A Gentleman in Moscow*, in his real-life role as a Russia correspondent. If you were hoping that he fished the bottle out of the Volga River, though, I will have to disappoint you. He picked it up on the beach at Vineyard Haven, approximately two miles from West Chop.

Set in Moscow's famous Hotel Metropol, *A Gentleman in Moscow* tells the story of thirty-two years in the life of Count Alexander Ilyich Rostov, sentenced in 1922 by a Bolshevik court to house arrest for life at the hotel, where he has been living. Rostov is a brilliant character—denounced for being an inveterate and unrepentant aristocrat, he makes a life for himself in the hotel over the decades of his enforced residence there, holed up in a sixth-floor attic, while life outside in Russia is changing beyond all recognition. His life is spared only because the committee that sentences him approves of a poem he wrote in 1913.

If you have read the book, you might have noticed that the date June 21 crops up repeatedly, and that several key events occur on this date over the years. This is just the tip of the iceberg in terms of the hidden structure, which Towles describes as accordion-like. Those thirty-two years over which the story takes place might have given you a hint that powers of 2 might be involved somehow (because 32 is 2^5, or $2 \times 2 \times 2 \times 2 \times 2$). And indeed they are. The book begins by recounting the events of June 21, 1922, the summer solstice, the day when Rostov begins his house arrest. We then hear what happens one day after the arrival at the hotel, then two days, and then five days. Then it's ten days after, three weeks, six weeks, three months, six months, and eventually we reach the precise anniversary, June 21, 1923. The time periods are (roughly) doubling each time. The doubling continues: we revisit Rostov on the summer solstice two years, four years, eight years, and finally sixteen years after his time at the Metropol begins—to 1938. This is the midpoint of the story, just as the summer solstice is the midpoint of the year, when the days are at their longest and the nights shortest. What Towles now does is to pivot on this midpoint and, in a lovely symmetry, to reverse the sequence—we jump eight years to 1946 (which is eight years before the end), then four years, two years and so on, repeatedly halving the intervals, until the concluding part of the book, which again occurs on June 21, the anniversary of the count's arrival at the Metropol. I won't tell you what happens, but it's a very pleasing conclusion.

Obviously, if you are like me, you may be feeling some discomfort at the claim that a sequence beginning 1, 2, 5, 10, 21 (three weeks) can properly be described as a doubling. After all, two plus two equals five only in the Orwellian torture chambers of *1984*. But it all works if you start with one year and then round down to the nearest sensible unit. Half of one year is six months; half of that is three months. Half of three months is six-and-a-bit weeks, so we round down to six weeks, and halving again is three weeks, or twenty-one days. Half of twenty-one days is ten days and change, half again is five days, half of five days is $2\frac{1}{2}$, so we round down to two, and then the final halving brings us to one day. I hereby give my professorial seal of approval!

Just as in *The Luminaries*, this choice of mathematical structure, a geometric progression and its inverse, is not random: it serves the narrative. At the beginning, that "granularity," as Towles refers to it, is necessary in order that we, and Rostov, can be properly introduced to the Hotel Metropol, his new attic quarters, and the other people who live there—the guests and staff. As time passes, it is more appropriate to travel more quickly—you wouldn't want to hear in detail about every day for thirty years. But this process should not continue indefinitely. As the end of the narrative approaches, we need that granularity again, leading up to the final and exciting conclusion (which I won't spoil for you). The doubling and shrinking is an excellent way to achieve this. It is also something like the way human memory works, and how we experience the passage of time. We all have very distinct memories of childhood, but then time seems to speed up during our adulthood; as we close in on the present moment, we have good recollection of today, yesterday, and the immediate past, but time contracts as our memories recede into the past.

Sequences of doublings and halvings occur along the number line, but for a two-dimensional example of mathematical structure in literature, we need look no further than Georges Perec's critically acclaimed *Life: A User's Manual* (or, to give it its original French title, *La Vie mode d'emploi*). As I mentioned earlier, all its action takes place at a precise moment in time. Having subverted any possibility of temporal structure, then, the way is open to impose some other framework. The story is set in an apartment building in Paris, 11 rue Simon-Crubellier, where the lives of its many inhabitants are intertwined in a profusion of ways. There's Bartlebooth, the eccentric Englishman who has spent years learning to paint and traveling the world creating watercolors of different ports, which he then has converted into jigsaw puzzles that he makes it his life's work to reassemble. The puzzle maker and Bartlebooth's painting tutor are fellow residents of 11 rue Simon-Crubellier. Unfortunately, Bartlebooth fails in his goal because he dies just before 8:00 p.m. on June 23, 1975, before he can complete all the puzzles.

The visible part of the structural edifice in this novel is the fact that the apartment building has 100 rooms, in a 10×10 square array. This includes attic rooms, basements, and stairwells, by the way. Each chapter is set in a different room. So far, so good. But the structure goes much deeper than that, and the story of the mathematics behind it involves card games, imperial Russia, early computers, and a mistake made by one of the world's greatest mathematicians.

Have you ever solved a sudoku? If you have, you have constructed what's known as a Latin square. If you haven't, don't worry; I've made a very small sudoku to show you what I mean. (This one is 4×4, but usually in newspapers they are 9×9.) The grid must end up with the numbers 1 through 4 each appearing exactly once in each row and exactly once in each column. I've filled in some of the numbers already; your job is to complete the grid so that every row and every column has a 1, a 2, a 3, and a 4. (If this were a 9×9 grid, we'd be working with the numbers 1 through 9.)

3	1		
		1	3
4	2		
1		2	

You can solve it by spotting, for instance, that the first column must contain a 2, so that gap in the first column must be a 2, and that forces the entry in the second row/second column to be 4, and so on. The completed grid is this:

3	1	4	2
2	4	1	3
4	2	3	1
1	3	2	4

A square grid like this with all the numbers appearing exactly once in each row and column is called a Latin square.

If you were a seventeenth-century French aristocrat looking for an entertaining logical diversion, you might have tried a different Latin square puzzle that was doing the rounds at the time. This one also involves a 4 × 4 grid, but this time it's made of playing cards. In the puzzle, you have to arrange the four highest cards (jack, queen, king, ace—we call these the "court cards" in the UK) in each of the four suits (hearts, diamonds, spades, clubs) of a deck of cards in a 4 × 4 grid, in a special way. Every row and column has to contain exactly one card from each suit, and every row and column has to contain four different court cards (an ace, a king, a queen, and a jack). One solution to this is shown.

A♠	K♥	Q♣	J♦
K♦	A♣	J♥	Q♠
J♣	Q♦	K♠	A♥
Q♥	J♠	A♦	K♣

What we end up with here is not just one Latin square, but two. There's a Latin square of suits and also one of card names. In addition to this, they play nicely with each other in that each combination is there exactly once—we don't have two queens of hearts, for instance. So this is a kind of "double" Latin square. Specifically, it's a pair of Latin squares involving two different sets of numbers or symbols, overlaid in such a way that every pair of symbols occurs exactly once. These are sometimes called "orthogonal Latin squares," or "bi-Latin squares," or "Greco-Latin squares"—the last of these because one set of symbols is taken from the Greek alphabet and the other from the Latin. But I'll stick with "double Latin square."

There are lots of different solutions to this card puzzle, but the precise number (1,152) wasn't known until the British mathematician Kathleen Ollerenshaw worked it out several centuries later. She was quite a woman. Born in 1912, she enjoyed mathematics very much as a child, and when she became deaf after an illness at age eight, she found that it was one of the few subjects (as they were taught at that time) in which her deafness did not impede her. During her long mathematical career she also produced the first academic paper setting forth a method to solve a Rubik's Cube from any starting position—this feat coming with the side effect of a thumb injury due to too much cube manipulation, an ailment described by *Reader's Digest* as the first known case of "mathematician's thumb." Oh, and she became lord mayor of Manchester and played ice hockey for England. Call me an old romantic, but I do like the fact that she married her childhood sweetheart, Robert Ollerenshaw, saying that she knew it must be love when he sent her a slide rule as a present.

Back to our card puzzles. The many available solutions would certainly be enough to entertain you for a few winter evenings. But after a while, a bigger challenge was needed, and just such a puzzle became popular in the 1770s. It's known as the "36 officers problem." This time, you have six different regiments, each with six officers of six different ranks—lieutenant, captain, major, and so on. Again, you have to put them into a square grid, 6×6 this time, so that there's exactly one of each rank and one of each regiment in each row and column. What you need is a 6×6 double Latin square. Now, this problem was doing the rounds of the St. Petersburg aristocracy, and the story goes that Catherine the Great, empress regnant of all Russia, no less, was intrigued by it; got stuck putting her colonels, brigadiers, and generals into position; and called in the hotshot mathematician Leonhard Euler, who was in Russia at the time at the St. Petersburg academy, to help. And here's the thing: Euler couldn't do it either.

Two things you need to know about Euler: first, his name is

pronounced "Oiler," and second, he is one of the most admired and influential mathematicians of all time, with ninety-two volumes of mathematical works to his name. He single-handedly initiated the mathematical research area known as graph theory, among multitudinous other things. He introduced a lot of our modern mathematical notation, including the way we write functions. The French mathematician Pierre-Simon Laplace, himself no slouch, exhorted us, "Read Euler, read Euler, he is the master of us all." So if Euler couldn't do something, then we pay attention. Like all mathematicians, when I can't solve a problem (and if that never happens to you, it means you aren't asking hard enough questions), I have to decide: Is it just me failing, or is this thing actually impossible? The next step is to crystallize this feeling with a conjecture: This problem has no solution—it is impossible. Of course, once you say this out loud, in a mathematical paper or at a conference, there's a chance that someone sees a way to do it, and then you feel a bit silly. So when you make a conjecture, you want to be pretty confident about it. And that's what Euler did with the 36 officers problem: he conjectured that it wasn't just him failing to spot a solution, but that *no solution is possible*—no 6 × 6 double Latin square exists.

The only way to be sure the thing really is impossible is to prove it mathematically. You have to give some reason why there cannot be a solution. Just to give you an idea, I can show you that the "4 officers problem" can't be solved—in other words, there does not exist a 2 × 2 double Latin square. This would have two ranks and two regiments. Let's say you have a general and a major from Regiment 1 and Regiment 2, and you have to put these four officers in a 2 × 2 square grid such that each row and column contains one major and one general, and one officer from each regiment. Since you can't have both generals in the same row or column, they must be at opposite corners, and there are only two ways to do this, as I've shown in this picture.

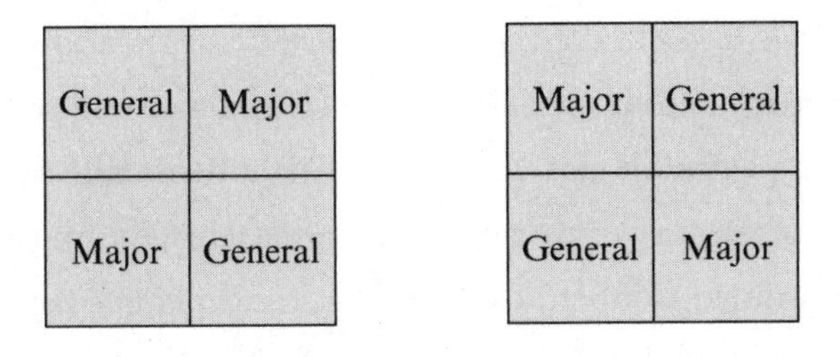

Whatever corner he's in, General 1 is therefore sharing a row with one major and a column with the other. But oh, no! This means there's either a row or a column with both General 1 and Major 1 in it, and that breaks the rule about having one officer from each regiment in each row and column. Which is a major problem (pun absolutely intended). Being a mathematician—we can't help ourselves—Euler then looked for a general (ha again!) rule about what sizes of double Latin square can exist. He knew that 2 × 2 is impossible, and that the next one he couldn't solve was 6 × 6. He went further and managed to prove that there's always a double Latin square for odd sizes (3 × 3, 5 × 5, and so on) and when the size is a multiple of four (4 × 4—the card puzzle—8 × 8, and so on). Euler conjectured in 1782 that for the remaining numbers, 2, 6, 10, 14, 18, and so on, going up in fours, a double Latin square of that size is impossible. Even to resolve the 6 × 6 case, the number of possibilities to rule out is astronomical: literally millions. It was finally done, in what mathematicians aptly call a "proof by exhaustion," by Gaston Tarry in 1901. He didn't check all the cases one by one—don't worry. He found some clever ways to rule out lots of batches of cases in one go, so reducing the problem to checking a smaller number, but still quite a lot, of possibilities. The upshot is that Catherine the Great and Euler the even greater were right: the 36 officers problem is impossible. So it was looking good for Euler's conjecture.

But then something amazing happened. In 1959, in one of the first uses of early computers in mathematics, E. T. Parker, R. C. Bose, and S. S. Shrikhande managed to find a 10 × 10 double Latin square! Even more amazingly, they showed that these squares exist for all other numbers bigger than 6, even the troublesome 14, 18, and so on. Euler had been wrong—but it took nearly two centuries to find out. This was big news—a picture of the "impossible" 10 × 10 double Latin square was shown on the cover of the November 1959 edition of *Scientific American*. And here's where we come back to Georges Perec. He was particularly interested in exploring the potential uses of mathematical structures to create new literary forms and constraints, so the exciting discovery of double Latin squares that Euler thought were impossible was right up Perec's street.

In *Life: A User's Manual*, the setting is a building with ten floors and ten rooms on each floor, creating a 10 × 10 square with a hundred rooms. Perec then created several lists of ten characteristics—a list of ten fabrics, for example. Each chapter of the novel takes place in a particular room in the building, and thus corresponds to a particular place in the 10 × 10 square. By overlaying the relevant double Latin squares, each chapter features unique combinations of different characteristics taken from these lists of ten. This results in a very rich narrative structure. And we're not done yet—there's a final component to add to this combinatorial fireworks display. The stories are told consecutively by following an order of rooms given by a knight's tour of a 10 × 10 chessboard. In chess (which is played on an 8 × 8 square), the knight is the only piece that cannot move to an adjacent square. It moves two spaces in one direction and then one in the perpendicular direction (for example, two spaces right and one space up). A knight's tour is accomplished when a knight travels over the entire chessboard, visiting each space exactly once. It's not obvious that this is possible, but it is, and the first recorded solution is attributed to one al-Adli ar-Rumi, who lived in Baghdad in the mid-ninth century. Are there other solutions? Is it possible for other size chessboards? The first systematic study of knight's tours was carried out by, you guessed it, Euler—and the answer to both questions is yes.

I should say that there's no shame in making a conjecture that turns out to be false. Euler's conjecture led to exciting mathematics and took centuries to resolve. So in a minute when I call him a failure, it's definitely tongue-in-cheek. I can only dream of being as successful a failure as Euler! One of the themes, then, of *Life: A User's Manual* is failure. Bartlebooth fails in his life's mission of assembling all those jigsaw puzzles. Valene, a painter who also lives in the apartment building, fails in his plan to paint a picture of the building showing all the rooms and the people in them. The double Latin square of the structure includes in its story Euler's failure to predict its existence. And there's one final failure, deliberately introduced by Perec: the knight fails to tour the building! The book has not 100 chapters, but 99—there's a basement room that is missed. I love Perec's "explanation," when describing the construction of the book and its failure to visit all the rooms: it's not his fault. "For this," he says, "the little girl on pages 295 and 394 is solely responsible."[5]

We have seen in this chapter how writers like Eleanor Catton and Amor Towles have exploited mathematical structures to powerful effect in shaping the chronologies of their novels. Georges Perec, in *Life: A User's Manual*, stepped things up considerably with intricate designs that combine both space and time in the geometry of his narrative. But Perec was just one member of the Oulipo, the group of writers I've mentioned before who do dazzling things at the frontiers of literary constraint. They are the subject of the next chapter.

3

A Workshop for Potential Literature

Mathematics and the Oulipo

On November 24, 1960, at a café in Paris, two Frenchmen, Raymond Queneau and François Le Lionnais, met with a group of fellow mathematically minded writers and literature-minded mathematicians and formed the Ouvroir de littérature potentielle, or Oulipo (from the first letters of the words). This translates roughly as "workshop of potential literature." The aim of the group was to explore new possibilities for structures that could be used in literature, whether that be poetry, novels, or plays. Since mathematics is the lodestone of structure, the group was particularly interested in how mathematical ideas could be starting points for new literary forms and structural constraints. Queneau and Le Lionnais are not well known outside literary circles, though you are now familiar with Queneau's 100 trillion sonnets, but you have probably encountered Oulipians like Italo Calvino and Marcel Duchamp—and we have already made friends with Georges Perec.

The question of how to make something new in art is hardly unique to 1960s France, nor is it unique to literature. The Oulipian response—use mathematics—can be seen as in some part a reaction to the surrealists, with their automatic writing and other techniques to get material out of the subconscious and onto the page. The basic idea of the Oulipo is that one way to create new kinds of literature is to create new literary forms and work within them. And what is a literary form but the imposition of some kind of constraint upon words—the number of lines in a sonnet, for example? Even the building blocks of language are usually understood to come with rules—a sentence "must" contain a noun and a verb, for instance.

There was something going on in mathematics writing at the time too that influenced the Oulipians, and that was the series of books by Nicolas Bourbaki that had been appearing at regular intervals since the 1940s. Here's an interesting thing about Bourbaki: there is not now, and never has been, a mathematician by that name. Bourbaki was the pseudonym of a group of mostly French mathematicians who got together to write, collectively and anonymously, a series of books that would cover what you might call the entire architectural foundation of modern mathematics, from first principles. It's a pretty amazing story—and these books are still in use today. The volume I have, covering one of my own research interests in algebra, is certainly rather dog-eared.

The practice of setting out the rules of engagement for your subject and then proving theorems based on this solid foundation has a noble lineage, going back thousands of years to Euclid. The rules of engagement are first to define the words you're going to use, just to establish that we all mean the same thing when we say "circle" or "line," and then to establish some starting points—things that we agree are true and from which we can deduce further truths. What this approach does for mathematicians is exactly what a constraint like the sonnet form does for poets: it gives you a structure and then invites you to explore it. What can I achieve within this setting? Within the rules of Euclidean geometry, we can prove Pythagoras's theorem. Within the rules of sonnets, we can write "Shall I compare thee to a summer's day? / Thou art more lovely and more temperate."

So what sort of "axioms" might make sense in literature? One very simple example is the one used in what are called *lipograms*. These are texts in which certain letters are forbidden. (The word "lipogram" comes from the ancient Greek for "leaving out a letter.") The best-known lipogrammatic novel is our friend Georges Perec's *La Disparition*, published in 1969. It is a text satisfying a single axiom: The letter *e* is forbidden. Now, in most, if not all, European languages, *e* is the most challenging letter to omit, because it is the most common. In French, more than one-sixth of the letters in normal text are *e*'s (including accented versions like *é* and *è* and *ê*). Try to write just one sentence without an *e*. It's difficult to do. (See what I did there? Or rather: look at my action just now.)

The Oulipo did not invent lipograms. They have a long history going back as far as ancient Greece, where the sixth-century BCE poet Lasus of Hermione wrote at least two poems that deliberately avoided using the letter sigma, apparently because he didn't like it. To each his own, I suppose. A tenth-century Byzantine encyclopedia called the *Suda* mentions a much more ambitious enterprise. It tells of a poet called Tryphiodorus who, nearly a thousand years after Lasus, produced a lipogrammatic version of Homer's *Odyssey*. The *Odyssey* has 24 books, and the Greek alphabet, at least at that time, had 24 letters. Each book of Tryphiodorus's *Odyssey* (which is now lost, sadly) omitted one letter—the first book had no α, the second had no β, and so on. *La Disparition* was not even the first novel to omit the letter *e*. That honor goes to *Gadsby*, a now almost forgotten 1939 novel by Ernest Vincent Wright. The Oulipians have a cheeky term for work produced in an Oulipian spirit that happens to predate the foundation of the Oulipo: anticipatory plagiarism. (There is a knowing wink to the anticipatory plagiarism of *Gadsby* in *La Disparition*—a character named Lord Gadsby V. Wright.)

With Wright's work, and every other lipogram, there is always a question: Yes, it's clever, but why do it? Does it help to make good art? There is no particular reason to write *Gadsby* without the letter *e*—nothing in the text makes that choice especially relevant. I have nothing against an intellectual challenge, but you want to feel that it's not just a sterile game. This, I think, is what raises *La Disparition* above almost all other lipogrammatic texts. It's not just that it's an incredibly difficult technical challenge to produce an entire book without the most common letter in your language. The added extra is that Perec's book follows one of two precepts set out by fellow Oulipian Jacques Roubaud: that a text written within a given constraint must in some way refer to the constraint. (I'll tell you about his second precept later.) The plot of *La Disparition* revolves around the search for something missing, which the characters in the novel eventually realize is the letter *e*. There are clues for the readers, such as chapters numbered 1 to 26 with Chapter 5 missing (because *e* is the fifth letter of the alphabet). But there are also clues for the characters in the novel. There is a hospital ward with 26 beds, but nobody is in bed 5. There's a 26-volume encyclopedia with no Volume V. Roubaud described *La Disparition* as "a novel about a disappearance, the disappearance of the *e*; it is thus both the story of what it recounts and the story of the constraint that creates that which is recounted."

In the novel's postscript, Perec explains, still without the letter *e*, why he has written it: "My ambition, as Author, my point, I would go so far as to say my fixation, my constant fixation, was primarily to concoct an artifact as original as it was illuminating, an artifact that would, or just possibly might, act as a stimulant on notions of construction, of narration, of plotting, of action, a stimulant, in a word, on fiction-writing today." (The *e*-less English translation here is by Gilbert Adair.) The literary critic and leading Perec authority Warren F. Motte has suggested that *La Disparition* is also a meditation on loss. Perec was orphaned in the Second World War—his father was killed in action and his mother was murdered in the Holocaust. As Motte points out, the absence of *e* means that "Perec cannot say the words *père*, *mère*, *parents*, *famille* in his novel, nor can he write the name Georges Perec. In short, each 'void' in the novel is abundantly furnished with meaning, and each points toward the existential void that Perec grappled with throughout his youth and early adulthood."

Which is harder to write: a novel with no *e* like Perec's *La Disparition* or his sequel *Les Revenentes*, whose *only* vowel is *e*? Raymond Queneau suggested a mathematical way to measure "lipogrammatic difficulty."

We all instinctively know that omission of the letter *x*, say, would be easier than omission of the letter *t*, and of course the longer such a text is, the harder it is to create. Queneau's idea was to give a precise measure of this difficulty using the frequency distribution of letters for the language in which your text is written. Any particular text will have slightly different proportions of the different letters, but if you collate results over large swaths of different texts, you can predict quite accurately the proportions of each letter that you'd expect to find in a piece of written English. The most common letters, in order, are *e*, *t*, *a*, *i*, and *o*. The least common are *z*, *q*, *x*, *j*, and *k*. This knowledge was used for centuries to crack secret codes, because if your adversary has encrypted a message by substituting letters for other letters or symbols, then you can make an educated guess that the most commonly occurring symbol represents the letter *e*, and the second most common is *t*, and so on (unless your adversary has written a lipogram). We'll have more to say on this in Chapter 8.

Queneau's measure of lipogrammatic difficulty is to take the frequency, *f*, of the letter or letters omitted and multiply by the length *n* of the text in words. In English, for instance, out of every hundred letters in a typical text, on average two of them would be *y*'s and thirteen would be *e*'s. This is because the frequency of *y* is 0.02, and the frequency of *e* is 0.13. So we predict $0.02 \times 100 = 2$ letter *y*'s, and $0.13 \times 100 = 13$ letter *e*'s. We can be more accurate if we want. To five decimal places, the frequency of *e* is 0.12702, and that of *y* is 0.01974.

Let's see how this works in practice. The lipogrammatic difficulty of creating a text of five hundred words with no letter *y* is $0.01974 \times 500 = 10$, rounded to the nearest whole number. But it's significantly harder to create even a much shorter text with no *e*. A two-hundred-word *e*-less text has difficulty level $0.12702 \times 200 = 25$ (again rounded to the nearest whole number). What about *La Disparition*? In French, *e* is even more common than in English, with a frequency of 0.16716. This gives an eighty-thousand-word novel like *La Disparition* an enormous difficulty of 13,373. Now, *La Disparition* was translated into English as *A Void* by Gilbert Adair. If one started from a blank slate, the difficulty level of an eighty-thousand-word *e*-less text in English would be 10,162. But please don't imagine for a second that I think writing *A Void* was easier than writing *La Disparition*. Translators like Adair have to face the formidable challenge of maintaining the lipogrammatic constraint while at the same time producing a faithful translation. It is an astonishing achievement.

A safer comparison is between the difficulty of *La Disparition* and the difficulty of *Les Revenentes*, Perec's follow-up novel that he joked used up all the *e*'s missing from *La Disparition*. This time, to work out the difficulty we have to add up the frequencies in French of all the other vowels together. I've checked this and come up with a total frequency of 0.28018; I've also done a very rough word count of *Les Revenentes* and arrived at a total of thirty-six thousand words—that means a difficulty level of 10,086. It's obvious why *Les Revenentes* is shorter. If it were the same length as *La Disparition*, it would be verging on twice as hard to write.[1]

A more recent lipogrammatic text that has the same self-referential quality as *La Disparition* is 2001's *Ella Minnow Pea* by Mark Dunn—there's a hint of what is to follow even in the title character's name, which sounds like the sequence *l*, *m*, *n*, *o*, *p*. The book is set on the fictional island of Nollop, whose inhabitants revere Nevin Nollop, putative inventor of the pangram "The quick brown fox jumps over the lazy dog." (In case you're unfamiliar, a pangram is a phrase or sentence featuring every letter of the alphabet.) There's a statue of Nollop on the island, with the pangram inscribed below it. One day, one of the tiles on which the letters of the pangram are written falls off, and the island's rulers take this as a sort of divine instruction that this letter must be stricken from the alphabet and banned. At this point, it disappears from the text in the book. As more letters fall from the statue, they are banned too. The only way for this process to cease, decides the government, is if it were to turn out that Nollop is not actually a deity—and that can be true only if a shorter pangram is found. At the most desperate moment, when only the letters *l*, *m*, *n*, *o*, and *p* remain, the eponymous Ella manages to find a thirty-two-letter pangram (three letters shorter than Nollop's), the full alphabet can be restored, and they all live happily ever after.

Before moving on from lipograms, I'll just mention *Eunoia*, by the Canadian author Christian Bök, which won Canada's Griffin Poetry Prize in 2002. There are five chapters in the main part of the book that each use only one vowel, with the letter *y* being omitted throughout. A sample sentence from Chapter A is "A law as harsh as a *fatwa* bans all paragraphs that lack an A as a standard hallmark." The title, *Eunoia*, is the shortest English word that contains all the vowels—it means a state of good health. The shortest such word in French is somewhat better known—*oiseau*, meaning "bird"—which is the title of the second part of the book. As the final section, "The New Ennui," states, the text "makes a Sisyphean spectacle of its labour, wilfully crippling its language in order to show that, even under such improbable conditions of duress, language can still express an uncanny, if not sublime, thought." This is beautifully put, and there's certainly some lovely imagery in the book. But I have to say that, though we can admire the extraordinary technical accomplishment required to produce such a lipogrammatic tour de force, the ratio of clever technique to emotional punch in the work is a little too high in places. With *Eunoia*, I think it's time to draw our discussion of lipograms to a close.

There is something so French, somehow, about the Oulipo—where else could it possibly have been formed than in a Parisian café? Yet probably the most famous Oulipian was a Cuban-born Italian. His mother gave her son a name intended to remind him of his heritage, only to move back to Italy almost immediately, meaning that Italo Calvino (for it is he) was saddled forever with a name that he described as "belligerently nationalistic."

Calvino's best-known work is *If on a Winter's Night a Traveler*, one of those rare books written in the second person. It's about a reader (you) trying to read a book called *If on a Winter's Night a Traveler*. You buy it, but it has the same sixteen pages repeated over and over again. When you return it, it turns out that actually these pages are copies of a different book called *Outside the Town of Malbork*. But something goes wrong when you try to find that book too, leaving you with the tantalizing start of a third book. Sections recounting your attempts to get hold of these books alternate with the beginning chapter of each. It's clever and funny and includes a list of book categories that is instantly recognizable to the inveterate book buyer (including Books You Could Put Aside Maybe to Read This Summer, Books You Want to Own So They'll Be Handy Just in Case, and Books That Everybody's Read So It's as If You Had Read Them Too).

Perhaps you have already put aside *If on a Winter's Night a Traveler*, maybe to read this summer, so let me persuade you also to look at Calvino's beautiful, melancholy book *Invisible Cities*. To use another of his categories, it is definitely a Book You Need to Go with Other Books on Your Shelves: *Invisible Cities* nods both to the travels of Marco Polo and to Thomas More's *Utopia*, with a hint of *One Thousand and One Arabian Nights*. The book contains fantastical descriptions of fifty-five cities that are supposed to have been in Kublai Khan's empire, ranging in length from a paragraph or two to a couple of pages. Argia, the underground city, for example, gets just fourteen lines—nothing can be seen from above ground, and it's hard to know whether the city is there at all. "The place is deserted," Calvino writes. "At night, putting your ear to the ground, you can sometimes hear a door slam." Behind all of these cities is the one city that Marco Polo never speaks about, but of which every other city is just a reflection: his home. "Every time I describe a city I am saying something about Venice," he tells the Khan.

Invisible Cities is divided into nine chapters, but the way the accounts of the cities are divided up and numbered is rather curious. Each city falls into a particular one of eleven categories (such as "Cities & the

Dead," or "Continuous Cities"), with five cities of each type. Chapter 2, for example, is shown like this in the table of contents:

2.

. . . . .

Cities & Memory • 5

Cities & Desire • 4

Cities & Signs • 3

Thin Cities • 2

Trading Cities • 1

. . . . .

The dotted lines represent unnamed sections in each chapter that contain conversations between Marco Polo and Kublai Khan. Chapters 3 through 8 also have five cities, numbered 5, 4, 3, 2, 1. But Chapter 1 and Chapter 9 have ten cities each, numbered seemingly (but in fact not) at random. Chapter 1 does not contain any 5s and Chapter 9 has no 1s. What is going on? Why this descending order 5, 4, 3, 2, 1? Why not just have eleven chapters with five cities in each, or five chapters with eleven? Why have fifty-five cities in the first place? Let's begin with that last question.

One of the inspirations for *Invisible Cities* was *Utopia*, by Thomas More. Thomas More was a Tudor writer and statesman, eventually becoming lord chancellor of England under Henry VIII. Unfortunately, he opposed Henry's decision to separate England from the Catholic Church and was executed for treason because of it. His 1516 book, *Utopia*, is the account of an imagined perfect country. ("Utopia," a word Thomas More coined, is derived from the Greek for either "no place" or "good place," depending how you convert that "U" into Greek.) Only one city, Amaurot, is described in detail because we are told that all the others are similar. So Calvino is filling in the gaps in More's work by telling us about all fifty-five cities.

Hold the front page, though. When I looked at English translations of

Utopia (it was written in Latin), they all said that there are 54 cities. This is rather curious. I don't know if Calvino owned an Italian translation in which there are erroneously 55 cities. Or are we to understand that there are 54 cities in addition to Amaurot? One edition has a footnote that the 54 cities of Utopia "parallel the fifty-three counties that made up England and Wales in More's time, plus one for London." My Latin's not up to much, but the original *"quatuor et quinquaginta"* does look awfully like 54, even to me. I don't want to start an international incident here, so to keep the peace let me suggest that if *Utopia*'s 54 comes from 53 + 1, perhaps *Invisible Cities* is 54 + 1, in tribute.

Now that we have 55 cities (somehow), how shall we arrange them into chapters? Well, we have eleven kinds of city, with five of each kind. So we could have a structure like a rectangle, with each row representing a chapter, and each column a type of city, like this. Here, the numbers 1, 2, 3, 4, and 5 are the five cities of each type. The first column is "Cities & Memory," the second is "Cities & Desire," and so on to the eleventh column, which is "Hidden Cities."

Chapter 1	1	1	1	1	1	1	1	1	1	1	1
Chapter 2	2	2	2	2	2	2	2	2	2	2	2
Chapter 3	3	3	3	3	3	3	3	3	3	3	3
Chapter 4	4	4	4	4	4	4	4	4	4	4	4
Chapter 5	5	5	5	5	5	5	5	5	5	5	5

Chapter 1 here has City 1 of each type; Chapter 2 has City 2; and so on. Let's not mince words: This structure is boring. Cycling through the same eleven elements in the same order each time does not give a feeling of progression and does not allow for different chapters to have different flavors.

A clue to the structure chosen by Calvino is given in the text. "My Empire," says Kublai Khan, "is made of the stuff of crystals, its molecules arranged in a perfect pattern. Amid the surge of the elements, a splendid hard diamond takes shape." What Calvino does is shift each of the columns successively downward like this:

1										
2	1									
3	2	1								
4	3	2	1							
5	4	3	2	1						
	5	4	3	2	1					
		5	4	3	2	1				
			5	4	3	2	1			
				5	4	3	2	1		
					5	4	3	2	1	
						5	4	3	2	1
							5	4	3	2
								5	4	3
									5	4
										5

In order to avoid chapters with just one or two cities in them, and to create a pleasing symmetry, Calvino gives Chapter 1 and Chapter 9 the first and last four rows of this structure. The intervening chapters now all have a 5, 4, 3, 2, 1 pattern, in which we see the fifth example of one type, then the fourth of the next, and so on. Each chapter, we visit a given type for the final time and introduce a new type. This mix of old and new, familiar and unfamiliar, gives a subtle momentum to the book's framework.

	1										
	2	1									
Chapter 1	3	2	1								
	4	3	2	1							
Chapter 2	5	4	3	2	1						
Chapter 3		5	4	3	2	1					
Chapter 4			5	4	3	2	1				
Chapter 5				5	4	3	2	1			
Chapter 6					5	4	3	2	1		
Chapter 7						5	4	3	2	1	
Chapter 8							5	4	3	2	1
								5	4	3	2
Chapter 9									5	4	3
										5	4
											5

We can also notice that if we take Chapter 1 and Chapter 9 together, they fit together to form a microcosm of the whole, with exactly four copies of the "5 4 3 2 1" motif. And there we are: "A splendid hard diamond takes shape." It is a very elegant design. Indeed, Calvino himself said that *Invisible Cities* was one of the works with which he was most pleased, because in it he managed to say the "maximum number of things in the smallest number of words."

There's one more Easter egg in the book's structure for the mathematical reader. In Chapter 8 of *Invisible Cities*, Kublai Khan meditates on the game of chess (of course it's in Chapter 8, because chess is played on an 8 × 8 square): "If each city is like a game of chess, the day when I have learned the rules, I shall finally possess my empire, even if I shall never succeed in knowing all the cities it contains." We have seen the patterns drawn in the book's structure, and look—55 cities plus 9 chapters equals the 64 squares on a chessboard. Coincidence? Not a chance.

Like any successful franchise, the Oulipo has several spin-offs. "Oulipo," remember, stands for "Ouvroir de littérature potentielle," or "workshop for potential literature." Any creative endeavor can have a "workshop for potential X," or "Ou-X-po." There's the Oubapo (bandes dessinées—comic strips), the Oupeinpo (peinture—painting), and even the Oulipopo, the Ouvroir de littérature policière potentielle: the workshop for potential detective fiction. There are many potential potential workshops. What's really needed, naturally, is an Ou-ou-X-po-po, and that would have to be followed by an Ou-ou-ou-X-po-po-po, and so on . . . but I digress.

If you like your Oulipo with a side order of murder, look no further than Claude Berge's *Qui a tué le Duc de Densmore?* (*Who Killed the Duke of Densmore?*). Berge was a respected French mathematician who made significant contributions to graph theory, and also a long-standing member of the Oulipo. He loved both mathematics and literature (a man after my own heart) and found it hard to decide which should be his career focus: "I wasn't quite sure that I wanted to do mathematics. There was often a greater urge to study literature." Berge's story about the murder of the Duke of Densmore not only uses a mathematical idea, it also uses a mathematical consequence of that idea. In this way it respects the second of the precepts put forward by Jacques Roubaud.

The first of these, if you recall, says that a text using a particular constraint must mention that constraint in some way. The second says that if a mathematical idea is used, then some consequence of that idea should also be incorporated. Berge's story involves a famous detective trying to solve an old case—the Duke of Densmore had been murdered years ago, but the culprit is still at large. The pool of suspects is narrowed down to a group of seven lady friends (to put it coyly) of the duke. Each of them visited the duke's house in the time leading up to the murder. Over the intervening years, they all claim to have forgotten the exact dates of their visits. But they do remember who else was there at the time. If two people met, then their visits must have coincided, if only briefly. What our detective ends up with, then, is a collection of intervals of time, and all he knows about them is which ones overlap.

This doesn't seem a lot to go on. But there's a clever way to visualize the connections in a situation like this—it's called an *interval graph*. Graphs in this sense of the word are something like a map of a subway system—you have various points (subway stations, or time intervals), and you join together the ones that are connected (adjacent stops on a subway line, or time intervals that overlap). For an example, let's take a literary family—the March girls from *Little Women*. Suppose that Meg, Jo, Beth, and Amy all visit their grumpy aunt. Meg says she saw Jo and Amy there, Jo says she saw Meg and Beth, Beth reports meeting Jo and Amy, and Amy sees Beth and Meg. All that information can be captured efficiently in this graph, which, if everyone is telling the truth, is an example of an interval graph:

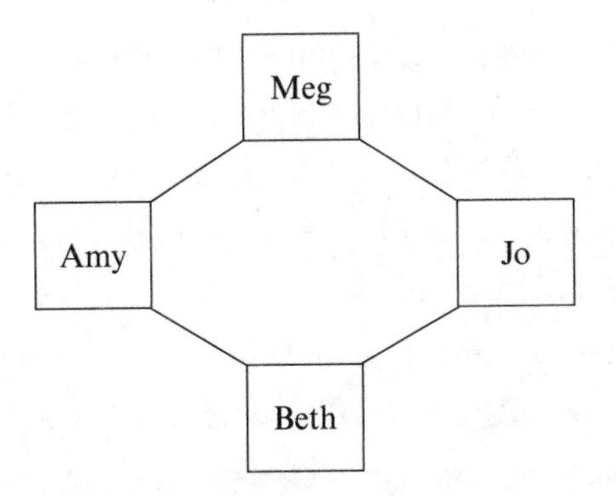

Here's the thing, though. This graph has a cycle Meg—Jo—Beth—Amy. But there is a theorem in graph theory that says that every interval graph is "chordal." What this means is that somewhere in every cycle there has to be a chord—an intermediate line joining two of its points. If a graph doesn't have that property, then it can't be a true interval graph. Here, it means there must be a line either joining Meg to Beth, or one joining Amy to Jo. The inescapable conclusion, though it pains me to say so, is that at least one of the March girls isn't telling the truth. Marmee will be so disappointed. With this example, we can't prove who is lying. (My money's on Amy.) But in the Densmore story, there are more suspects, and the graph has the property that there is exactly one person who, if we removed them from the graph, leaves the rest of the graph as a real interval graph. And what better reason to lie than that you murdered the duke? The detective knows the interval graph theorem and catches the killer.

As we have seen throughout this chapter, the Oulipian approach is at once playful and earnest—one of my favorite combinations. Life, as they say, is too important to be taken seriously. To close our guided tour of all things Oulipo, I've been inspired to invent my own piece of potential literature. I don't think it has been done before, but if it has, then I congratulate my predecessors on their excellent bit of anticipatory plagiarism.

In 1976, Raymond Queneau published a short article titled "Foundations of Literature (after David Hilbert)." David Hilbert was an important nineteenth- and twentieth-century mathematician who did a lot to put mathematics, and especially geometry, on a firm, rigorous footing. In geometry, people had spent the best part of two thousand years trying to sort out once and for all what the hell was going on with Euclid's parallel postulate, the axiom that says if you start with a line and take any point not on that line, there is exactly one line through that point that is parallel to the line you started with. Nobody could prove this, which is why it had to be taken as an axiom. But it's less obvious than the other axioms. What people realized in the nineteenth century is that in fact there are versions of geometry—so-called non-Euclidean geometries—in which the parallel postulate actually doesn't hold, meaning that some of the properties of Euclidean geometry may no longer follow. For example, take planet Earth. Draw a triangle by heading from the North Pole down to the equator, then traveling a quarter of the way along the equator, and then going back to the North Pole. This triangle has three right angles! Have we just destroyed geometry? No. What's happened is that we have discovered that the geometry of curved surfaces is different from that of flat surfaces. Here's another example—perspective drawing. In perspective drawing, parallel lines meet at a "vanishing point." This is a bit of a downer if your definition of "parallel" is "never meeting."

What Hilbert did, brilliantly, was to set up some rules, or axioms, of geometry that would be general enough to cover all these different examples, and many others, while keeping the things they all have in common. Here are two of Hilbert's axioms:

1. Given two distinct points, there is always a line containing those points.
2. Given a line, any two points on the line uniquely determine that line.

Together, these rules say that any two points will lie on one, and only one, line. These axioms are true in standard geometry, but they are also true for the curved "lines" on a sphere and for the lines in a perspective drawing. In fact, there are lots of situations in which there is a useful concept of "lines" and "points." The important insight is that as long as the axioms are true for our particular setup, however weird and wacky, then all the consequences of those axioms will also be true. So we can prove theorems that are true in a bunch of different scenarios, with no extra effort.

Back to "Foundations of Literature," then. Queneau suggests that literary texts could be created subject to specific literary axioms. Instead of points and lines, we could talk about words and sentences. Having created a set of axioms, your new literary form will consist of texts that satisfy those axioms. The two geometric axioms we had earlier, says Queneau, would then become the following:

1. Given two distinct words in the text, there is always a sentence in the text containing those two words.
2. Given a sentence in the text, any two words in the text uniquely determine that sentence.

As Queneau points out, the text describing the axioms does not itself satisfy the axioms, and that's fine—the definition of (say) a rhyming couplet is not necessarily itself a rhyming couplet, though naturally I now want to think of one that is.

Let me show you a truly strange "geometry." It's called the Fano plane, named for the Italian mathematician Gino Fano, who discovered it. (In fact, at least two other anticipatory plagiarists had independently discovered it before him, though I don't think he was aware of this.) The Fano plane contains precisely seven points, and precisely seven "lines"—in my picture they are shown as six straight lines and a circle. Each line consists of exactly three points.

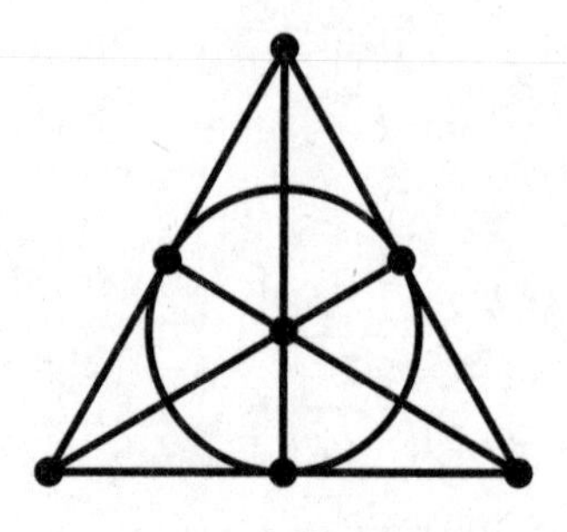

The Fano Plane

This object is breathtakingly symmetrical. Every pair of points lies on exactly one line and every pair of lines meets in exactly one point. Every line contains precisely three points and every point lies on precisely three lines. It's beautiful. Yes, there are about a million applications of this structure, everything from cryptography to lottery tickets, from set theory to experiment design. There's also a link to a picture that may be more familiar—the classic Venn diagram showing all the possible intersections between three sets: the seven regions of the diagram correspond to the seven points of the Fano plane. But the reason I love the Fano plane has nothing to do with the applications. It's purely down to its symmetrical simplicity.

As a tribute to Queneau and the Oulipo, I have created a new axiomatic literary form, which I have christened "Fano fiction." The rules for Fano fiction are simple. Each text uses a vocabulary of exactly seven words (our "points") and consists of exactly seven sentences (our "lines"), each of which contains exactly three words. Each pair of words appears in exactly one sentence, and any pair of sentences has exactly one word in common. I've also required of myself that each sentence should observe the traditional grammatical rule of having a verb in it. With only twenty-one words in total, it's going to have to be a pretty spare narrative. My inaugural work of Fano fiction is encapsulated in the Fano plane diagram on the following page.

The story tells how you, a talent agency employee, were advised that it's best to get hold of the next big talent, and book her fast. A T-shirt

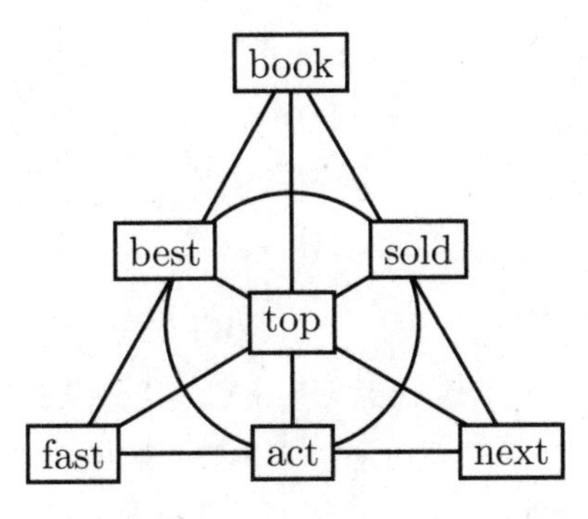

line she endorsed flew off the shelves, and there was a bidding war for her autobiography. You encouraged her to write the follow-up volume without delay, and to top her best previous achievements. She did so well that you could sell your share of the proceeds and retire a millionaire. And here's the Fano fiction version:

"Book top act!
Best book fast!"
Top sold fast.
Next, book sold.
"Act fast—next!
Next: top best!"
Best act: sold.

All I need to do now is sit back and wait for my Nobel Prize in Literature.

I said at the end of the last chapter that the members of the Oulipo take the use of constraints to extremes. Do they go too far down that route? One accusation sometimes leveled at their work is that the constraints imposed serve only to create clever puzzles. The first response to this objection is that there's no reason something can't be clever and great art at the same time. But more important—and this is a point Oulipians themselves have reminded critics of from time to time—the Oulipo is a workshop of *potential* literature. Its purpose is to provide possible structures, not necessarily to provide the literature itself. As Raymond Queneau said, "We place ourselves beyond aesthetic value, which does not mean that we despise it."

The fact that many terrible sonnets have been written in the course of history does not imply that the concept of the sonnet is inherently bad; there is some boring, arid constrained writing, just as there are boring, arid novels. But there are also fantastic, imaginative, creative, exciting works of constrained writing—Perec, Calvino, Queneau, and others have produced art that we are still talking about. So that's my defense of the Oulipo. Everyone will have their own personal sense of the boundary between art and artifice, but I truly think there's something Oulipian to suit every taste.

4

Let Me Count the Ways

The Arithmetic of Narrative Choice

Have you ever played one of those story apps on your phone that require you to make a choice at the end of each "scene"? I can almost hear the brain cells dissolving as my daughter decides whether her character should go to the prom with Chad or with Kyle. Naturally, I can't help but wonder how many ways through such games there are, and how many scenes have to be written. Many books, plays, and even poems give us a choice of how to read them. Mathematics can help us to understand the implications. Imagine an interactive story in which at the end of each page you pick one of two options, each of which takes you to a different page. On the face of it, you'd need two different second pages, but then four different third pages, eight fourth pages, and so on. Incredibly, even if you make only ten choices in the whole book, it would need to be more than two thousand pages long! This obviously can't be how such books are constructed.

In this chapter, we'll look at the mathematics of narrative choice. We'll learn how to write a play in which the audience gets to decide what happens next without the actors' having to learn hundreds of scenes, and we'll explore what happens when you write a story in the shape of a Möbius strip.

. . .

We saw in Chapter 2 some playful examples of graphs representing the plots of stories. But there is a different kind of graph that can be used in plays, books, or other forms of literature in which the creators make available more than one path through the text. This can be done by directing the reader in various ways, or by giving the reader (or theater-goer) choices at key points, or by introducing randomness. The graphs I'm talking about are networks with points, or vertices, joined by edges that represent some sort of link between the points, like the interval graphs I showed you in the last chapter. The example I gave there was a subway map. For these kinds of maps, what we care about is the connections, not exact distances or accurate geographical location. Another graph that's very important in today's world is one in which each vertex is a web page, and we join two pages when one includes a link to the other. These kinds of graphs represent the connectedness of the Internet and help to determine how highly pages are ranked in search engines. Pages with lots of links are higher up the list. Finally, if you have ever played Six Degrees of Kevin Bacon, you'll know that we can also represent the connectedness of society with a graph in which every person is a vertex and two people are linked if they have appeared in (or directed or otherwise been involved with) the same movie.

I'm now going to show you a graph that was devised by Oulipo member Paul Fournel, along with Jean-Pierre Enard. It's called a *theater tree*, and it was created to help write interactive plays. The idea is that at the end of each scene, the actors ask the audience to choose between two possible plot developments. A masked man walks onto the stage at the end of a scene, say. The audience are asked: Is this man the king's illegitimate son, or is he the queen's lover? The audience's choice determines the scene to be played next. This is fun for the audience, but think about the poor actors (not to mention the poor set builders, costume designers, and prop wranglers): every time there is another choice, the number of scenes the actors have to learn goes up, and it goes up dramatically. If the audience makes four choices in total, then they will see five scenes (a choice at the end of Scenes 1 to 4, and then the final Scene 5). But how many scenes do the actors have to learn? There's one Scene 1. Then the audience makes a choice, and so there would be two versions of Scene 2. Then there's another choice, so these two Scene 2s bifurcate into four Scene 3s, then eight Scene 4s, and finally sixteen Scene 5s. If you add this up, you get thirty-one scenes. (Is this $2^5 - 1$? Yes, it is.) The scene structure can be shown in a graph, working down from Scene 1 at the top to all the possible Scene 5s at the bottom:

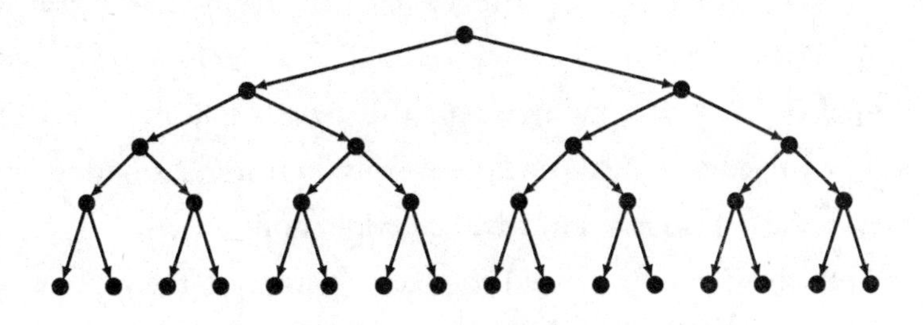

Here's where Fournel and Enard come in. They used the fact that there are other graphs that start from a single vertex (Scene 1) and still have a choice of two paths below each vertex but have fewer vertices in total. This means that the audience can still experience an interactive play with five choices, but the actors will be a lot happier.

Let's look at Fournel and Enard's suggested graph—the theater tree. As you can see, there are only fifteen scenes:

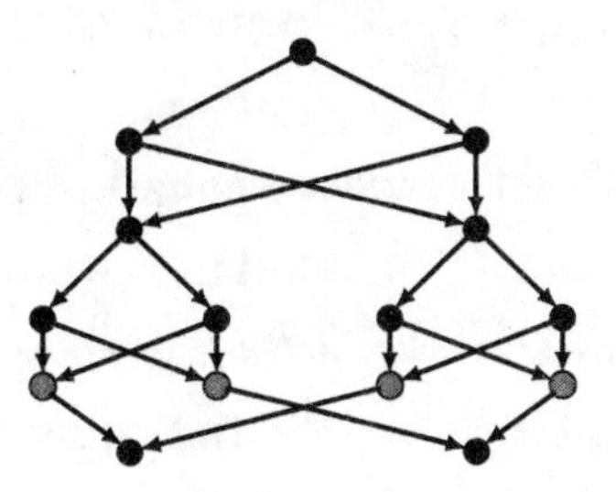

How does this work? In the theater tree, we start at the top with Scene 1, then work down, making our choice at each point. But whereas in the original play we had four different versions of Scene 3, the theater tree manages to have just two. How? Well, the writer has to ensure that whichever version of Scene 2 is played, by the end of the scene it's possible to choose between the same two options for Scene 3. For example, if the audience decided at the end of Scene 1 that the masked stranger was the king's son, then in their Scene 2 a new person, the queen's lover, could dramatically appear. Meanwhile, for those who said the stranger should be the queen's lover, their Scene 2 could introduce the king's son. This way, in either case, the decision at the end of Scene 2 can be "Should the king's son and the queen's lover fight a duel, or should they turn out to be old friends?" At the end of Scene 4, the audience again is given a choice: "Do you want a happy ending, or a tragic ending?" In order for the four Scene 4s to resolve to just two endings, the final scene is split into a bridging scene—5a (in gray)—and a conclusion—5b (in black). Counting up all the scenes and half scenes, we arrive at a total of fifteen.

In both these setups, the audience sees five scenes. How many possible plays are there for the audience? The answer is the total number of paths. In both cases, the answer is sixteen. This is because each choice splits the play universe in two. With one choice, there are two plays. With two choices, you get four plays; with three choices it's eight, and with four choices it's sixteen. Paul Fournel pointed out that creating sixteen separate five-scene plays would require writing eighty scenes. Even the inefficient version of an interactive play improved on this with thirty-one scenes. But the theater tree is even better. Graph theory has reduced the performers' workload by sixty-five scenes: an impressive 81 percent.

I wondered whether it was possible to do even better. The answer is yes, but with a caveat. If you went to see an interactive play created using the theater tree, your experience—four choices made, five scenes—would be indistinguishable from one created using the larger 31-scene tree. At least, it would be if you saw it only once. If you went back the next day for another go, there's a good chance you'd see some of the same scenes, even if you picked different options. The illusion of the full tree works for only a single viewing—which is fine, of course. During that single viewing you would not feel that your choices were illusory in any way. However, here is a more efficient tree that still gives the audience four choices:

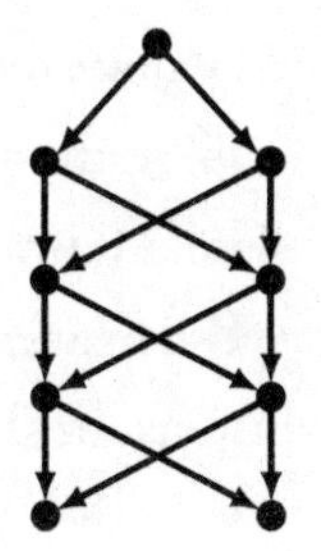

This time, I've deployed the same trick that the theater tree uses to prevent the need for four different Scene 3s, but I've applied it to every scene. It requires that each time a choice is made, by the end of the resulting scene the choice has to turn out to have been irrelevant. Should two people fight or turn out to be friends? The scene that follows must somehow make both things true, whichever the audience picked, so that subsequent choices are not constrained. This is because at each point, only two scenes are available, so both have to make sense, whatever the sequence of choices. If the audience votes for a fight, then maybe the characters start fighting, but it emerges that they are friends who have had a disagreement. If the audience wants them to be friends, then fine, they greet each other as friends but an argument ensues that results in a fight. It would likely become fairly obvious as an audience member in a play like this that your choices are not really having any effect. My graph may be more efficient, but it will likely lead to a worse play than a theater tree would.

I'm not aware of vast numbers of productions of theater-tree plays. But there have certainly been interactive TV shows, one example being 2018's *Bandersnatch*, part of the *Black Mirror* series on Netflix. It has 150 minutes of footage ingeniously put together to create 250 segments. Choices made by the viewer determine which segments are played, and in what order. There are reportedly more than a trillion paths through the story, each lasting on average ninety minutes. Such shows would be prohibitively expensive to make unless efficient graphs are used. Without them, every choice doubles the number of scenes that must be written and filmed, so that, as I said at the start of this chapter, even ten choices over eleven scenes would require more than two thousand scenes (the exact number is $2^{11} - 1 = 2047$). The analogue of the hyperefficient but boring graph in which there are just two versions of each scene after the first would require twenty-one scenes—but viewers would soon smell a rat. The best solutions will be somewhere between these two extremes.

There is a form of literature that uses just this kind of combination of free choice and hidden structure, but on a vastly more ambitious scale. I'm talking about the "choose your own adventure"–style books that many of us had as kids. They were very popular in the 1980s, then fell out of favor as computer games started to be able to create the same kind of experience. But now they are having a bit of a revival. If you're not familiar with this genre, basically you, the reader, are a character in the book: you are thrown into events, and you have to decide what to do at regular intervals. If you want to investigate that mysterious cave you just found, turn to page 144. If you want to take the path to the castle, turn to page 81. If instead you want to cross the bridge and fight the troll, turn to page 121. Sometimes randomness is introduced—you may need to roll dice to determine whether you beat the troll in the fight you decided to pick with it. If you win, go to page 94; if you lose, go to page 26. Reading these books involves potentially hundreds of choices (unless you foolishly choose to get into an argument with a troll, in which case you're likely a goner after only a few pages). The arithmetic of choice tells us straightaway that many pages must appear on several narrative paths. Otherwise, in a book with 100 choices, even if there are just two options each time, the book would be $2^{101} - 1$ pages long. Even if each page were just a tenth of a millimeter thick, it would take the light from your flashlight (I assume you are reading the book under the covers because your parents told you to go to sleep hours ago) 26.8 billion years to get from the first page to the last. In reality, a scaled-up version of something like the theater tree has to be used.

To find out more, I needed to talk to an expert. For my ninth birthday, I was given a book called *The Warlock of Firetop Mountain*, the first volume in the wildly successful *Fighting Fantasy* series of interactive books in which "You are the hero."[1] The book came out in 1982, and it was written by Ian Livingstone and Steve Jackson. Since these names have been etched on my brain for over forty years, I was delighted when Sir Ian, as he is now known, agreed to speak to me about how he constructs his branching narrative adventure stories. As well as being the co-creator of the *Fighting Fantasy* series, which has sold more than twenty million books worldwide, he's also a gaming legend—co-founder of Games Workshop, which brought Dungeons & Dragons to the UK, and of Eidos Interactive, publisher of the Tomb Raider games.

When we meet, Sir Ian explains that he creates each book using a flowchart, done by hand, and he shows me the original chart for *Deathtrap Dungeon* (Puffin, 1984). He starts with a basic path and then gradually adds branch points—places where decisions are made. Very little is predetermined. "We know the overall story arc, but what happens along the way is an iterative process." For example, "You might decide you want an iron door, then you think, 'Well, how do we get in here? Is it open? No, I want it to be locked, there's something important in there.' So we need a key. . . . You then go back earlier in the story and add a box to a room that they've been in, and the key is in the box." Each event or decision is numbered at random, and those numbers are then crossed off a master list (the *Fighting Fantasy* books all have four hundred sections, or "references"). There are many story strands in play, but there are always what Sir Ian calls *pinch points*, at which you go back to a node that gives important information and brings the story back into one passageway again. These pinch points are vital in preventing an exponential increase in the number of possible choices.

As you go along with the writing, you have to keep checking that there is definitely at least one successful path through the book, as well as making sure there aren't any loops from which there's no escape. And then there's the question of difficulty. There's great skill in designing the book so that the challenge is at the right level—not enough monsters to fight, and it's too easy; too many, and the reader becomes dispirited. "Oh, no, not *another* army of the undead," you sigh. Sir Ian's books are carefully calibrated to avoid either extreme. He does have fun with readers, though. "My joy is always trying to lure people to their doom," he jokes. "The petals along the floor where they fall on the poison spikes." He also enjoys the occasional red herring, "littering the dungeon with useless objects that they pick up, and then they miss the important items."

At that point I think guiltily of the decisions I'd advised my daughter Emma to make earlier that day when we'd been reading *The Warlock of Firetop Mountain* together. "So you're saying that maybe the reader goes straight for the shiny silver amulet, but actually . . ."

"It's the wooden duck you needed, yes," he says.

So be warned.

There are differences between constructing a game book and constructing a computer game. In a computer game, the program can keep track of what objects are where. Suppose one section says, "You enter the hidden chamber. There is a bag of gold on the floor, which you may take if you wish. You may exit either north or east." In a computer game, if you take the bag of gold, then if you go back to the chamber, the program will not tell you there is a bag of gold on the floor. But the book can't tell whether or not you have picked up the gold without having two versions of the rest of the story, which would double the length of the rest of the book. It's also very clunky to have instructions like "This room has some gold in it, unless you already took the gold on a previous visit." So the book can't allow you to return to that room.

If you can't go backward and forward like this, how many choices *would* a reader get to make, and therefore how many parts of the book will they see, during a typical read-through? It's usually between 100 and 150, says Sir Ian. For me that's a very impressive ratio—you are seeing around a third of the book's content each time, while making a very large number of choices.

That leads to another crucial property of the book's design: a single choice must not cut out huge swaths of the adventure, or there won't be room for those 150 choices. Remember that the writer also has to maintain the overall story arc to make it a compelling adventure. Each choice has to be meaningful too. There has to be an actual consequence of going left instead of right, or of talking to a person or not talking to them. "Because if it's the same no matter what you choose, then why bother making it interactive? There are multiple layers of the component parts of making a thrilling adventure in which you are the hero."

There's a real mathematical tension here between efficiency, control, and choice. We've already seen that to avoid a book the size of a house, we have to have pinch points where many story lines converge. This means that great skill is required in the wording of these passages. Readers will be coming from several points, and whatever happens has to make sense to everyone. There's another aspect of word choice that Steve Jackson and Ian Livingstone were very careful with, right from the start: "I'm proud to say that we never assumed it was a male playing these books. . . . When they meet someone it's 'Fellow stranger' and 'You're a very fine-looking person.' . . . I'm very proud that even in 1982 we did that, and I think that's key to its popularity." I think he's right.

Finally—asking for a friend—what is Sir Ian's view on cheating? Good news: he's okay with it. "I call it peeking around the corner," he says. Another tactic is the "five-finger bookmark." This is the technique of keeping your fingers in the pages from the last few choices, so that if your latest decision proves to have been unwise, you can think better of it. Discretion, after all, is the better part of valor.

In "choose your own adventure" books, readers influence their own journey through the story. But even when the author remains fully in the driver's seat, the narrative path down which they direct us may be far from a straight line. The simplest examples are what are called *reverse poems*. These poems are first read in the normal way, from top to bottom, but then after the last line the reader is asked to reread the poem in reverse, from bottom to top. Usually, the top-to-bottom version is pessimistic, and the bottom-to-top version challenges that negative worldview. The poem "Lost Generation," by Jonathan Reed, begins with these three lines:

I am part of a lost generation
And I refuse to believe that
I can change the world

When read in reverse, those first three lines become an optimistic assertion of possibility:

I can change the world
And I refuse to believe that
I am part of a lost generation

If you want to write your own reverse poem, there are plenty of templates available.[2] The way to do it is to have statements like "It's a fact that" or "It's not true that" interspersed with assertions, as in the following:

Math is just numbers.
It's not true that
Math is beautiful.

Now read it in reverse.

Geometrically speaking, what a reverse poem does is to add a mirror line so that we reflect the poem back on itself, creating a poetical palindrome. A more explicit use of geometry is found in the (very!) short story "Frame-Tale," by the American writer John Barth, which appears in his 1968 anthology, *Lost in the Funhouse*. A frame tale is a story within a story, like the play within a play in *Hamlet*. "Frame-Tale" consists of a single page, with a few words printed on each side, along with the instructions "Cut on dotted line, twist end once, and fasten AB to ab, CD to cd." Cutting along this line gives you a narrow strip. On the first side of the strip are the words ONCE UPON A TIME THERE. On the other side are the words WAS A STORY THAT BEGAN. Now, if you just glued the ends of the strip together you'd get a band with ONCE UPON A TIME THERE on the outside and WAS A STORY THAT BEGAN on the inside. But introducing the twist creates not a band but a mathematical surface known as a Möbius strip.

The Möbius strip is a strange and interesting thing. Discovered in 1858 by the German mathematician August Ferdinand Möbius, it has what sounds like an impossible property: it is a thing you can create from an ordinary piece of paper, but it has only one side. I beg you to make one right now. Just take a narrow strip of paper, give it a twist, and tape the ends together. If you hold the Möbius strip anywhere, one of your fingers is on the top side and one on the bottom. But if you draw a line along the center of the strip, starting on your chosen "top" side and parallel to the edges, you'll find the line eventually passing along what was the "bottom," and a bit later coming back to the place you started. What this means is that the Möbius strip has just one surface! In spite of this, it's still true that at any given point, there's a matching point on the reverse, so at each stage there appears to be a back and a front—but this is just an illusion. I can't stop myself from asking you to cut the Möbius strip along the central line you've just drawn and see what happens. Nothing to do with literature, but it's really cool. And if you cut the resulting thing in half along *its* center line, something even crazier happens—do try it.

Anyway, the effect of the instructions in Barth's story is to create an infinite loop of stories: "Once upon a time there was a story that began 'Once upon a time there was a story that began "Once upon a time there was a story that began 'Once upon a time . . .'"'" Here's the thing, though: the fact of its being a Möbius strip (a literal, physical plot twist) isn't really taken advantage of. The effect—of a story whose end is its own beginning, forming an endless loop—would be produced more simply with a circle. Just write the sentence "Once upon a time there was a story that began" on one side of a strip of paper and glue its ends together. So I would say that "Frame-Tale" really ought to be classified as a circular story, rather than a Möbius strip.

The best circular story I have read is by the Argentine novelist Julio Cortázar. "Continuity of Parks" is just over a single page long, so I hope you will forgive me for potentially spoiling it for you by summarizing the plot. A man sits down in the green chair in his study to finish reading a novel. In the novel, two lovers are planning a murder. After their final tryst, they depart into the night, she in one direction, he in the other. He silently enters the house of the man he plans to kill, creeps up the stairs, and enters the study, where his victim is sitting in his green chair, reading . . . And then you can of course start the story again and read it, this time knowing the fate of the man in the green chair.

In circular stories, every time we return to the beginning, every new "Once upon a time" adds another layer of narrative distance. If we use Hilbert Schenck's idea from Chapter 2, that each additional level of narrative distance creates another dimension in the story, then these circular stories are examples of infinite dimensional narratives. However, we can never actually realize these dimensions, because we must at some point put the story down. I don't know what the highest-dimensional story ever written is, and in some sense this is a battle that can't be won because as soon as we did find a winner, we could create a story beginning "I once read the following story" and then quote the now second-place tale in full.[3]

Circling back (if I may) to Möbius strips, at least one writer has made fuller use of their properties. The British author Gabriel Josipovici published a collection called *Mobius the Stripper* in 1974. (This isn't a typo, by the way—the spelling used by Josipovici is Mobius, not Möbius.) The title story has text split into a top and bottom half throughout. You can read either half first. The story in the top half is about a man called Mobius, who is indeed a stripper at a nightclub—he physically strips in order to try to mentally strip away the baggage of society and find his true self. The story in the bottom half is of a writer in a slump, trying to free his mind and come up with new ideas. A friend suggests he go and see the act of this guy Mobius, and that starts the writer thinking. He decides to make up a story about Mobius, even though he has never met him—that's where the bottom story ends. Now we can seamlessly loop back into the first story, but this time we see it as a story created by the writer.

This could have been just another circular story. However, Josipovici is cleverer than that. In traveling around a real Möbius strip, as we noticed earlier, at any point on your journey there is a corresponding point on the reverse side—you'll reach it exactly halfway through your trip. *Mobius the Stripper* mirrors this: events in the two halves of the story leak through into the other halves, just as ink on a Möbius strip would show through faintly on the other side. The stories bleed into each other, and it is impossible to say which is the "real" story—is the author writing a fictional account of a real Mobius, or is Mobius entirely imagined, in which case, where did the author get the idea from? Incidentally, there is a higher-dimensional analogue of a Möbius strip—a "solid" that doesn't have an inside or an outside. It's called a Klein bottle (after the mathematician Felix Klein). Please write to me if you've heard of any Klein-bottle-shaped novels!

The reader can choose from two possible paths through *Mobius the Stripper*, and many more through *The Warlock of Firetop Mountain*. But in all the examples we have seen so far, although readers can make choices, they are still following a road map created by the author. Even those 100 trillion sonnets from Chapter 1 require you to put the lines in a prescribed order. There are books, however, that throw away the map completely. Our combinatorial cavalcade continues with my contender for the best value purchase of all time, a 1969 book by the English writer B. S. Johnson, memorably described by his biographer Jonathan

Coe as "Britain's one-man literary avant-garde of the 1960s."[4] Johnson, born in London in 1933, was a fascinating character. His father was a stock clerk at a bookshop; his mother had been a maid and then a waitress. He didn't follow the sort of path we expect of our literary greats. By age fourteen he was at a school whose aim was to prepare pupils for future office work, where he was taught "shorthand, typing, commerce and book-keeping, besides the usual things." He left at seventeen with the School Certificate, which, in theory at least, qualified him to go to university, but "no one had ever gone to University from Kingston Day Commercial School." So he got a job.

Five years later, a friend at work (he was an accounts clerk in the payroll department of a bakery) showed him the prospectus for Birkbeck—a college of the University of London that held all its lectures in the evenings so that people working during the day could still get a university education. Birkbeck started in 1823 and is still going—I was amazed and delighted to discover Johnson's Birkbeck connection because I have been teaching there for nearly twenty years and am constantly banging on about the vital importance of giving people the chance to pursue higher education at any stage in their lives. Anyway, Johnson applied, was accepted, and started studying at Birkbeck in the autumn of 1955. He did well and decided to become a full-time student, transferring at age twenty-three to another London college, King's (in spite of the Birkbeck registrar's attempt to dissuade him by saying that at King's he would be "surrounded by eighteen-year-old girls"). He wrote poetry, plays, and film and television scripts, as well as soccer and tennis match reports for national newspapers, but it is for his seven novels that he is best remembered.

Each of them experiments with form. For example, in *Albert Angelo*, a hole is cut into pages 147 and 149 so that the reader can look ahead to an event that will take place on page 151—we can perhaps think of this as adding a loop to the "graph" of the story. In *House Mother Normal*, a story is told from nine different viewpoints over nine chapters, each, except the last, having twenty-one pages. But there is additional structure. Each event in the narrative occurs at exactly the same place on the same page of each chapter. The story then becomes, instead of a single line, a series of parallel curves overlaid—it is a plane rather than a line. Poignantly, the narrators at each stage have increasingly advanced forms of dementia, and as their thoughts become more fragmentary and disordered, this externally imposed structure becomes more or less the last remnant of order staving off the chaos of senility.

Johnson was not the first to experiment with this kind of structure. *House Mother Normal* echoes a 1947 short novel by Philip Toynbee, *Tea with Mrs. Goodman*, which features events described by seven characters entering and leaving the same room at various times, with, for example, Time Period 4 being described by Narrator C on page C4. But there is little humanity in *Tea with Mrs. Goodman*—it's another example of the fact that structure for the sake of structure, in literature just as in mathematics, risks being arid and pointless. As Jonathan Coe writes, "Everything that is sterile and academic in Toynbee's novel he [Johnson] humanizes: formal experiment becomes not a substitute for emotion and sympathetic involvement but the very means by which these things are brought about."

In 1969, B. S. Johnson published *The Unfortunates*. It is "a book in a box," consisting of twenty-seven chapters, or sections. The first and last chapter are specified, but there are also twenty-five intervening sections that can be read in any order. They are not numbered, and because the sections are not bound into a book, there is no default order to follow. Your path is totally random. Each reading order will give you a different experience because of the knowledge that you have, or don't have, when you read a given part of the plot. *The Unfortunates* was not the first book to exploit random choice. A few years earlier, the French writer Marc Saporta had published *Composition No. 1*, an unbound novel whose pages could be read in any order at all. But this makes it extraordinarily difficult to tell any kind of story, and moreover, it detracts from the randomness because, as Johnson wrote, it imposes a different kind of structure on the material, "another arbitrary unit—the page and what type can be fitted on it."

What turns *The Unfortunates* from an arch intellectual exercise into a successful and meaningful work of fiction is that the form is chosen for a reason, and the use of the form enhances the meaning of the work. The novel concerns a sports journalist traveling to report on a football (or, as our American friends have it, soccer) match. This arises from a real-life incident in Johnson's life, when as one of the sports reporters for *The Observer* newspaper, he was by chance assigned a match in Nottingham to report on. When he arrived at the train station, he realized with a jolt that this was the same town where he had first met a dear friend of his, Tony Tillinghast, who had recently died of cancer at just twenty-nine years old. Johnson described how on that day "the memories of Tony and the routine football reporting, the past and the present, interwove in a completely random manner, without chronology." When submitting the finished manuscript, Johnson wrote to his editor that "to me, at least, it really does reflect the random way in which past and present interact in the mind: it is an enactment of randomness which the bound book simply cannot achieve."

Each of us reading *The Unfortunates* constructs, by our choices, a different book. How many potential books, then, are in the *Unfortunates* box? As you might imagine, it's quite a lot! Let's do a toy example just to get a feel for things. Consider the obscure art house movie *The Incredibles*. If you are not familiar with it, this was a 2004 Pixar movie about a family of superheroes: Mr. Incredible, his wife, Elastigirl, and their kids, who also have various superpowers. Given how much money it and its 2015 sequel made, it was surely only a matter of time before we got origin story movies for Mr. Incredible and Elastigirl. In fact, having looked into it, I discovered that there was an official Disney book in 2018 called *A Real Stretch: An Elastigirl Prequel Story*. Let us postulate a future in which you can plan a movie marathon consisting of *The Incredibles* along with *Mr. Incredible: The Prequel* and *Elastigirl:*

The Prequel. The order in which you watch them will affect your experience of each film. How many *Incredibles* movie trilogy experiences can you have? Movie 1 can be any of the three. For Movie 2, you've already used up one option, so you now have only two movies to choose from. For Movie 3, you've used up two of the three options, so there's only one choice left. We can see the possibilities in a diagram:

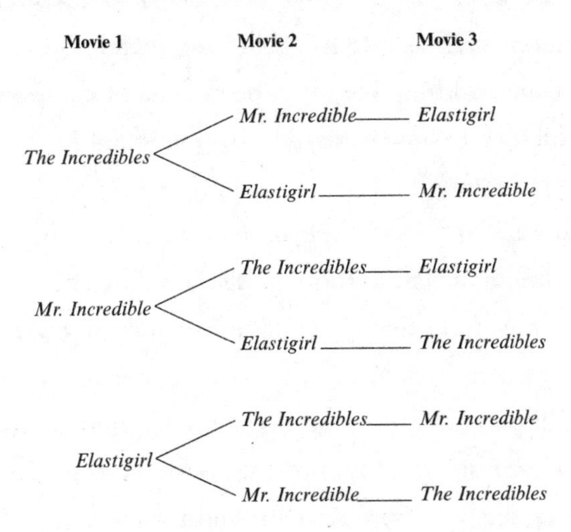

At each stage, the number of choices goes down by one. The total number of possible trilogies is $3 \times 2 \times 1 = 6$.

Okay, so now we've had a warm-up. It's time to go back from *The Incredibles* to *The Unfortunates* (poor things). In this case, the first and last chapters are fixed for us, and we read the middle 25 chapters in any order we like. This means there are 25 choices for your second chapter, 24 for your third (you've already used up one choice), 23 for your fourth, and so on, until there is just one possible 26th chapter left. The total number of ways to read the book is therefore

$$25 \times 24 \times 23 \times \cdots \times 2 \times 1$$

Mathematicians have a shorthand for this calculation—we write it as 25! to save ink (the exclamation point is read as "factorial"). In general, the factorial of a number N is the product of everything up to that number. So $3! = 3 \times 2 \times 1$, as we saw. The number $N!$ is the number of ways of ordering N things, and as N grows, $N!$ gets to be very large very fast. If you do all those multiplications to find 25!, you find that $25! = 25 \times 24 \times \cdots \times 2 \times 1 = 15{,}511{,}210{,}043{,}330{,}985{,}984{,}000{,}000$.

That's 15.5 septillion, if it helps (and I know it doesn't). If all eight billion people in the world dropped everything and each started reading a different version of *The Unfortunates* every day, it would take more than five trillion years for the full collection to be read. Going by my book club, some of whose members don't even manage our one book a month (ladies, you know who you are), I'm afraid to say we may not have time to get through it.

In case any defenders of *Composition No. 1* want to protest that it should win the "Best Value Book of All Time" award, as it has more potential versions, well, that's true. It consists of 150 pages, which can be read in any order. This means there are (to use our swanky factorial notation) 150! potential readings of the book, which is inconceivably many. To the nearest round number, it's 6 followed by 262 zeros. But this division of the book into so many short fragments really damages the quality of the narrative and, in my view at least, makes for a vastly inferior read. I have taken account of that in my careful consideration of the award, and I stand by my view that the prize should go to B. S. Johnson.

In the lacuna between reading book chapters in a random order and reading them from first to last lies Julio Cortázar's experimental novel *Hopscotch*. Cortázar was one of the most innovative writers of the last century, known for short stories such as "Blow-Up" (which inspired the 1966 Michelangelo Antonioni movie of the same name) and the circular story "Continuity of Parks" that I mentioned earlier. *Hopscotch* (*Rayuela* in Spanish) revolves around Horacio Oliveira, a discontented Argentinian intellectual, and the ragtag group of bohemians he associates with, especially his lover, La Maga, and Morelli, a novelist (and

Cortázar's alter ego), who himself is writing a novel that will be "out of line, untied, incongruous, minutely antinovelistic (although not anti-novelish)." The shape of the book is like the game of hopscotch, in which you alternately have your feet on the left, on the right, and in the middle. The book has 155 chapters. Chapters 1 to 36 are "From the Other Side," Chapters 37 to 56 are "From This Side," and Chapters 57 to 155 are "From Diverse Sides"—with a subtitle, "Expendable Chapters." We are encouraged to "play" the book—to hop from chapter to chapter, to be an active participant in the unfolding story.

Cortázar includes a page of instructions with two routes through the book. As he says, "In its own way, this book consists of many books, but two books above all." The first book is obtained by reading in a normal fashion, starting with Chapter 1 and working through the successive chapters in order. This book ends with Chapter 56, "at the close of which there are three garish little stars which stand for the words *The End.* Consequently, the reader may ignore what follows with a clean conscience." Of course, he doesn't really want you to do that, or not only that. He wants you to follow the second, more interesting path, which I'll explain in a moment. The book has a huge range of cultural references, but in particular, it contains many subtle and not-so-subtle allusions to that other great meandering narrative, *Tristram Shandy*, which I mentioned in Chapter 2.

One such echo is the annoying categorization of two forms of reader—the pedant who just reads chapters in order and then stops, versus the creative reader who joins in the fun—as the female reader and the male reader. Looking at my copy of *Tristram Shandy*, I found that Chapter 20 begins by asking the female reader to go back and read again the previous chapter, as she has missed an important point about Shandy's mother. While she goes back and looks again, Shandy tells us remaining readers,

> *I have imposed this penance upon the lady, neither out of wantonness nor cruelty; but from the best of motives; and therefore shall*

make her no apology for it when she returns back:—'Tis to rebuke a vicious taste, which has crept into thousands besides herself,—of reading straight forwards, more in quest of the adventures, than of the deep erudition and knowledge which a book of this cast, if read over as it should be, would infallibly impart with them—The mind should be accustomed to make wise reflections, and draw curious conclusions as it goes along.

Cortázar, similarly, is quoted as saying, "In *Hopscotch* I defined and attacked the Lady Reader who is incapable of waging true amorous battle against the book, a battle like that of Job with the angel." I can perhaps forgive such nomenclature from Laurence Sterne (1714–1768), but it's hard to stomach coming from Julio Cortázar (1914–1984).

The "hopscotch" route through *Hopscotch* is explained by Cortázar in his introduction. It should begin with Chapter 73 "and then [follow] the sequence indicated at the end of each chapter. In case of confusion or forgetfulness, one need only consult the following list," which list begins 73–1–2–116–3 and so on, interspersing one or more of the main 1–56 chapters, in order, with one or more of the expendable chapters. Whichever way you choose to read the novel, you will miss something. The "straight" reading gives you a story, but you don't get the two hundred pages of "Expendable Chapters": the footnotes, the digressions, the newspaper stories. The "hopscotch" route appears to cover everything, but brilliantly it actually misses out a complete chapter—Chapter 55. (I won't tell anyone if you cheat and read it anyway.) Also, if you actually follow the instructions, you will never finish reading the book. At the end of Chapter 77, you have read every chapter except 55, 58, and 131. Chapter 77 sends you to Chapter 131. Chapter 131 sends you to Chapter 58. Chapter 58 sends you to Chapter 131. Cortázar has trapped you in an infinite loop! He has created a paradox—a book that is both finite and infinite at the same time. If you follow his instructions, that is. Of course, truly inventive readers will refuse to abide even by Cortázar's rules and will choose their own way to engage with the book.

. . .

Assuming you have chosen to engage with this book by starting at the beginning and reading the chapters in order, then you'll have seen many ways that mathematics can shine a light on the hidden structures of literature. Next time you read a poem, you'll know how its patterns and rhythms have an underlying mathematical story to tell. You now understand how the choices you make in reading a book, and the choices the author makes in writing it, have mathematical implications for the shape and size of the narrative. Along the way I've shown you the strange and wonderful world of the Oulipo. You know how to quantify difficulty in lipograms, and how to construct 100 trillion stanzas[5] starting with just ten. Above all, I hope I've shown that behind every work of literature there is structure, and behind every structure there is delightful mathematics to explore.

Part II

Algebraic Allusions

The Narrative Uses of Mathematics

5

Fairy-Tale Figures

The Symbolism of Number in Fiction

Why do wishes come in threes? Why is it the seventh son of a seventh son who has magical powers? A handful of numbers—3, 7, 12, and 40 among them—seem particularly resonant, featuring in everything from religious texts to fairy tales, proverbs to nursery rhymes. In a very unscientific sample of numbers in phrase and fable, I inspected the contents of my own brain and came up with, among other things, Macbeth's three witches; Snow White's seven dwarfs; the three fates, three graces, and nine muses of ancient Greece; the nine realms of Norse mythology; the Five Pillars of Islam; and biblical references like the seven deadly sins, the twelve apostles, the twelve tribes of Israel, the forty days and nights of Noah's flood, the seventh seal, and so on. Some numbers have acquired more than their fair share of symbolic or cultural meaning. Is this just coincidence, or is there anything special, mathematically, about these chosen few magic numbers? I want to persuade you that, at least in part, there is.

In Part I, we looked at how mathematics can appear in the underlying structures of literature. Continuing our metaphor of the house of literature, Part II concerns how mathematics can furnish this house. The words themselves, the metaphors, the figures of speech: mathematics is to be found in all of them. I'll begin in this chapter with the most easily spotted manifestation of math: the use of numbers themselves. (I'll get on to Tolstoy's calculus metaphors later.)

Why do some numbers have more cultural significance and feature more in literature than others? This is a challenging question for mathematicians—the problem is that all numbers are our friends (something my sister, then aged five, said to our mother before going over to the dark side and defecting to physics). If a mathematician looks at any number hard enough, we can't help but find interesting things about it. I was being profiled for a British magazine called *Oh Comely* a few years ago, and since I'm a mathematician, and it was issue number 22 of the magazine, they asked me if I could tell them something interesting about the number 22. To begin with, I wasn't sure—it's not prime, it's not square. And actually at some point I gave up and decided I'd just talk about a fun mathematical puzzle I'd come across recently, which asks you to say what the next number in the following sequence is: 1, 11, 21, 1211, 111221 . . . you can try to guess the next number if you like before you read on. This sequence is called a "say what you see" sequence. Each term is just the description of the term before it. Beginning with "1," that's "one 1," so the next term is 11, and then 11 is "two 1's," so we write 21. Then 21 is "one 2, one 1," and we write 1211, and we continue in the same way with 111221 and so on. You can make a sequence like this with any number you like as the starting point.[1] Believe it or not, there is one, and only one, number, out of the infinitude of all numbers, that is fixed by this process. That is, if you "say what you see," you get the same number back. Guess what that number is? Yep, it's 22. That excellent coincidence is my long-winded way of reminding us that all numbers are interesting if you give them a chance.

Let's get back to our discussion of magic numbers, known to anthropologists as *pattern* numbers. Each of the smaller pattern numbers has its own distinctive personality, and different cultures have different favorites—though I'd argue that small odd numbers, in particular 3 and 7, seem to have the widest-ranging cultural resonances. The higher pattern numbers, on the other hand, are not chosen because of their individual character but rather tend to fall into one of three (again with the three) types. Probably your best bet if you want to be special is to be a round number, like 10, 12, 40 (40 has additional layers of symbolism that we'll talk about later), 100, or 1,000. These numbers, particularly the higher powers of 10, are not meant to be a literal count. (If I've told you that once, I've told you a hundred times.) Rather, they signify a generic large quantity. In Ireland you are greeted with a hundred thousand welcomes—*céad míle fáilte*. The Chinese version of the English birthday greeting "Many happy returns" is "May you live a hundred years." In the animal kingdom, what in English are known as "centipedes" in fact have just forty-two legs. By contrast, the same creature in German has a thousand legs (*Tausendfüßler*), whereas the Russian сороконожка (*sorokonozhka*) comes in closest to reality with forty legs (сорок being the Russian for "forty").

The second way for a big number to gain special significance is to be a kind of extrapolation of a smaller magic number. In the Old Testament book of Genesis, we are told that if Cain is avenged seven times, then Lamech shall be avenged seventy-seven times. (In another extrapolation, Lamech lives for 777 years.) There are also many occurrences of 70 and 7×70 in the Bible.

The third way for a large number to get into the picture is for it to be close to a round number. Numbers like 99 and 999 feel to us like upper limits—they are the biggest you can get without overstepping the boundary of the next big number. (This is why retailers use the psychological trick of pricing things at 99¢ or $9.99.) In the Islamic faith, according to one of the hadiths of Abu Hurayrah, Allah has ninety-nine names—that is, one hundred minus one—and whoever knows them will go to Paradise. In some traditions these ninety-nine names point to one most superior, greatest hundredth name (in Sufism it is "I am," for example). By contrast, numbers that are just above a large round number serve to emphasize their great size. Think of the thousand and one Arabian nights, or the long time conveyed by the expression "a year and a day," which is often how long it takes for brave heroes to return from adventures in fairy tales. Or even the "*mille tre*" (1,003) lovers that Leporello reports Don Giovanni to have had in the famous aria from Mozart's opera. And that's in Spain alone!

The large round numbers and their near neighbors relate to our counting system, which is based on the number 10. The reason for this is not mathematical, but anatomical: 10 is how high we can count before running out of fingers. Some cultures have counted in base 5 (one hand) or in base 20 (fingers and toes), and you see the occasional vestige of this in language—three score and ten (another biblical 70), or the French word for 99—"*quatre-vingt-dix-neuf*," which literally means "four times twenty plus ten plus nine." A wily six-appendaged female Martian might tell stories not for 1,001 nights ($10^3 + 1$) but 217 ($6^3 + 1$).

The two large numbers that seem not quite to fit in our list are 40 and 12. A dozen is a very useful quantity, and there's a mathematical reason for it, namely that it has lots of factors. The number 12 is divisible by 1, 2, 3, 4, and 6, so it's easy to share a dozen of something among a few people. In old predecimal English coinage, a shilling was twelve pence. That means you can easily make half a shilling (sixpence), a third of a shilling (fourpence, or one groat), a quarter of a shilling (thruppence), and one sixth of a shilling (tuppence). By the way, if you'll allow me to get something off my chest here: Wizard money in the Harry Potter books is the most irritatingly mathematically implausible currency in fiction. It just could not have emerged in the organic way that currency systems develop. There are three denominations: twenty-nine bronze Knuts make up a silver Sickle, and seventeen Sickles make a golden Galleon. Since 29 and 17 are both prime numbers, they can't be divided at all—you can't even have half a Galleon. How nonsensical!

Even though we Muggles nowadays use the decimal system for our money, we still buy our eggs in dozens, and we still divide our year into twelve months, with four three-month seasons, and our clock into twelve hours. Our ancient length measure, the foot, is twelve inches long. How long is an inch? That's easy: the English king Edward II defined it in 1324 to be the length of "3 barleycorns, round and dry." I don't keep up to date with trends in cobbling, so I'm unsure if it's still true, as King Edward decreed, that the difference between one shoe size and the next is the length of one barleycorn. The cultural significance of twelve includes the twelve apostles, the twelve days of Christmas, and fairy-tale dozens like the twelve princes who are turned into ravens in the Brothers Grimm tale "The Twelve Brothers." You may also recall the twelve dancing princesses who are carried in twelve boats across a magical lake each night to dance until dawn with twelve princes.

Just as 12 is a "good" number, 13, being 12 plus an odd 1 out, becomes a bad number by association. There were twelve apostles plus Jesus at the last supper, and we all know how that went. In our house, though, we like the number 13: not only were both my husband and my daughter born on the thirteenth day of the month, but for at least three years in a row, owing to the girls' passion for Taylor Swift, I've had to make a birthday cake for her on December 13. Meanwhile the 12 + 1 aspect of 13 gives us the baker's dozen. This phrase appears to have originated from a time when the law required bakers to sell goods like bread rolls by the dozen, but subject to a minimum weight requirement. In order not to risk coming in under the required weight, the baker would often throw in an extra roll just to be sure.

The number 40 is an interesting one. It has significant cultural resonance, which makes itself felt everywhere from "Ali Baba and the Forty Thieves" to the forty days and nights Jesus spent in the desert, which themselves mirror the forty days and nights Moses spent on Mount Sinai. If we take a nap, we say we are having forty winks; everybody in England knows that if it rains on Saint Swithin's Day, we will have forty days of rain; and of course we mustn't forget those forty-legged Russian centipedes. More topically, the origin of our word "quarantine" comes from the forty (*quaranta*) days that medieval visitors to Venice had to be kept in isolation to prevent the spread of plague. (Totally by coincidence, "40" also happens to be the answer to the famous quiz question

"What is the only number in the English language whose letters are in alphabetical order?")

Yes, 40 is "round" in the sense that it is a multiple of 10, but that's not enough of an explanation, since numbers like 30 and 50 are not accorded the same significance. There are a couple of things in 40's favor, though. First, it is "round" not only with respect to base 10, but also with respect to a base 20 counting system, being two score. One explanation in the context of time periods is that it's close to being 42, and 42 days is six weeks. But perhaps the true reason in this case is not mathematical but biological. I've personally been hyperaware of counting to 40 twice in my life because pregnancy lasts forty weeks. This could be one reason to associate the number 40 with a period of preparation ending in a great change.

Let's talk about small pattern numbers. You may remember that I began this chapter by sieving my brain for numbers, which resulted in a lot of instances of 3 and 7, with a few 5s and 9s thrown in. Does this mean that all small pattern numbers are odd? Well, no. If it means anything, it means that the contents of my brain are odd, but my family and friends could have told you that straightaway.

All of us are the products of our own heritage, but a fascinating picture emerges from exploring folktales and legends from a broader range of traditions. There are several cultures in which the numbers 4, 6, and 8 play important roles. I want to explore these even numbers first before moving on to odd numbers, culminating with 3 because I believe that of all the small pattern numbers it has the most far-reaching influence on the structure of narrative.

The number 4 occurs almost nowhere in European folktales, though it has a few cameo roles in English-language fiction: think T. S. Eliot's *Four Quartets*, John Updike's Rabbit Angstrom tetralogy, and the *Seasonal Quartet* by the Scottish author Ali Smith, a series of four interlinked novels named for the four seasons. In children's literature you might remember the four Pevensie children, Peter, Susan, Lucy, and Edmund, in C. S. Lewis's Narnia books, and the four Houses (Gryffindor, Ravenclaw, Hufflepuff, and Slytherin) at Hogwarts. There are obvious parallels between these, which I can't be the first to notice—not least the categorization of all of humanity as either brave, clever, kind, or evil. Aslan the lion, the god of Narnia, explicitly associates the Pevensie children with points of the compass (Peter is north, for example), which handily illustrates one reason why the number four occurs as a sacred number in all four corners (see what I did there?) of the globe. It is particularly prevalent in Native American creation stories—a claim that, naturally, I shall illustrate with four examples.

In Sioux and Lakota creation stories, the Creating Power remade the world by singing four songs. The first song caused rain to begin. With the second song the rain intensified. The third song caused the rivers to overflow. Then he sang a fourth song and stamped on the earth four times, splitting it open so the water covered the entire world and all the creatures from the old world died. Then he sent four animals to swim down through the waters to bring up a lump of mud. The loon, the otter, and the beaver all failed, but the turtle succeeded. This mud the Creating Power shaped into new lands. Then he made men and women from four colors of earth: red, white, black, and yellow.

A Chelan story, meanwhile, tells of four Wolf brothers, armed respectively with a one-forked, two-forked, three-forked, and four-forked spear, who killed the Great Beaver and divided its flesh into pieces from which different tribes were created. The Cherokee describe the earth as a huge floating island, held above the seas by four ropes representing the four sacred directions. Finally, in the Navajo tradition, we live in the fourth world, above three underworlds where the animal, insect, and spirit peoples dwell. When people arrived in the fourth world—our world—they named four sacred mountains and four sacred stones to lie at the borders of their lands. Ever Changing Woman, the wife of the Sun, made four clans from the flakes of her skin, and these are the descendants of the Diné, now known as the Navajo. My favorite part of the story comes when the Navajo gods are arranging the heavens, having placed the four sacred mountains in their proper positions. They put the sun and the moon in the sky, then start arranging the stars in a carefully designed pattern. But Coyote grows bored with waiting and tugs on the blanket where the stars are lying, hurling them randomly into the sky. That's why, even though deities prefer order, the stars are strewn chaotically around the heavens.

The four compass points give us a way to navigate the surface of the world. If we add in up and down, we get the six directions required to navigate in the air. But since humans have not been able to fly until rather recently (it didn't work out so well for Icarus), the number 6 has fewer cultural resonances than the number 4. Judeo-Christian tradition has it that the world was created in six days, the seventh day, the Sabbath, being when God put his feet up and had a rest, hence our seven-day week. The Qur'an also tells of creation in six stages, but the stages are usually taken to represent significant time periods—eons rather than days. The number 6 has the nice mathematical property that it is both the sum and the product of the first three numbers, because $6 = 1 + 2 + 3 = 1 \times 2 \times 3$. More than this, it was viewed by early mystics as being "perfect" because 6 is divisible by 1, 2, and 3, and so 6 is the sum of all of its factors. (I should really say "proper factors" because strictly speaking, 6 is a factor of itself.) This means that 6 is beautifully and exactly constructed from its own building blocks. Saint Augustine said that this was why God had chosen to create the world in six days, and that the structure of the creation was split precisely into $1 + 2 + 3$, with "Let there be light" on the first day, then two days to create the earth and the sea, followed by three days to create all the living things. The next perfect number after 6 is 28, because $28 = 1 + 2 + 4 + 7 + 14$. The Hellenistic Jewish philosopher Philo of Alexandria wrote that not only was the world created in six days because 6 is perfect, but this is also the reason why the lunar month is 28 days long.

If I had a mean streak, I'd suggest that you get a pencil and paper and try to find the next three perfect numbers. It would take a while.

The next one after 6 and 28 is 496, followed by 8,128, and then a gigantic gap until 33,550,336. I'm not aware of any appearances of these latter numbers in theology. Perfect numbers have been known, studied, and hunted for at least two thousand years. They are remarkably rare. At the time of writing, only fifty-one perfect numbers are known—we don't know whether there are more;[2] the last one to be found was discovered in 2018. So 6 is, mathematically speaking, very special and rare.

But in terms of literature, not many fairy tales use 6 as an important number, and I think that when they do the role of 6 is usually better understood as "7 minus 1" than as 6 in its own right. There are several German folktales featuring seven children: one sister and six brothers. I've also seen stories with twelve children: one sister and eleven brothers. I think it makes more sense to see these as 7s and 12s than 6s and 11s, though folklorists are welcome to send me their dissertations on the use of 6 in myth and legend.

In Chinese tradition, some numbers acquire connotations of good or bad luck by an accident of language, because sounds have different meanings based on intonation. The word for "eight" sounds like the word for "prosper," so is considered very auspicious, and efforts are made to include the number 8 in important events. The opening ceremony of the Beijing Olympics, for example, began at 8 minutes and 8 seconds past 8 p.m. on the 8th day of the 8th month of 2008. The word for "four," by contrast, sounds very similar to the word for "death," and so 4 is naturally considered unlucky (bad news for me, born on the fourth of April). But these are linguistic rather than mathematical considerations.

In cultures that do not have the added linguistic imperative to view 8 as lucky, it does still sometimes make an appearance. The Persian poet Amir Khusrau's twelfth-century work *Hasht-Bihisht* (eight paradises) is named for the traditional concept of the afterlife with eight paradises (one more than the seven hells, because God is merciful) surrounded by eight gates. This work is not well known in the English-speaking world, but you are likely to be familiar with a word derived from one of its tales: "The Three Princes of Serendip." (Serendip is the classical Persian name for Sri Lanka.) When the English writer Horace Walpole wanted a word to describe a fortunate chance event, he recalled this tale in which the princes were "always making discoveries, by accidents and sagacity, of things which they were not in quest of." And that's how, in 1754, the English language acquired the word "serendipity."

I have not yet mentioned the simplest numbers of all, the number 1 and the number 2. These are so instrumental that we almost don't see them. It would be barmy to assert that "Beauty and the Beast" is full of the number 1 because there is one beauty, one beast, one castle, one enchantress, one rose, and one talking teapot. The number 1 is somehow apart from all other numbers. This is true in mathematics as well. Even though a prime number is defined to be a number that can't be divided up into smaller factors (so that 3 is prime because we can write it only as 3×1 or 1×3, but 6 isn't prime because we can split it into 2×3 or 3×2), we exclude 1 from the list of primes. But it is the building block of all other numbers. We can make every number, or every whole number at least, by just adding 1 to itself enough times. It's the start of everything. But at the same time, if you have one of an item, you aren't really "counting" anything.

It's a little bit the same with 2. Even though it's incredibly important (the first and only even prime number, for a start), there's not enough going on with it for it to be a pattern number. A narrative with only binaries does not excite us for long. That being said, almost all fairy tales have at least one binary division: good versus evil, for example, Snow White versus the Evil Queen (#NotAllStepmothers). In mathematics, the number 2, being the first even number, is the first one that can be broken into two equal parts like this. Binary arithmetic, in which every number is expressed in terms of 1s and 0s (or true/false, or good/evil, if you prefer), is the basis of all computers. (It's also the basis of the old joke that there are 10 kinds of people in the world: those who understand binary and those who don't.)[3]

. . .

The fact that even numbers can be broken in half and also split into pairs could, I think, be a contributing factor to their having a slightly different role from odd numbers when they occur as small pattern numbers. The numbers 3, 5, and 7 are particularly "strong" in the sense that they can't be broken up. Not only are they odd, which means they can't be broken into two halves or split into pairs, but they are prime, so they can't be split up at all. The number 9, by contrast, isn't prime, but the only way to break it up is as three 3s, which makes it potentially extra-special, if 3 is a pattern number in your culture already. Shakespeare uses the number 9 in *Macbeth* as a magnifier of 3 in this way. The three witches, who make three prophecies and hail Macbeth by three titles (thane of Glamis, thane of Cawdor, and "king hereafter"), are a satanic version of the Holy Trinity. Here is what they chant as they circle the fire:

> The weird sisters, hand in hand,
> Posters of the sea and land,
> Thus do go, about, about:
> Thrice to thine, and thrice to mine
> And thrice again, to make up nine.

Nine can be even further magnified. In Act I, Scene 3, the first witch plans to curse a sailor whose wife has insulted her: "Weary se'nnights nine times nine / Shall he dwindle, peak and pine." In other words, the hex will last eighty-one weeks.

Nine can also be deployed in a similar way to 99 and 999—it is almost, but not quite, a large round number. This usage is sometimes found in Chinese folktales. For instance, in "The Bird with Nine Heads," the bird kidnaps a princess. When her rescuer comes to the cave where she is imprisoned, he sees her tending to the bird's wound, "for the hound of heaven had bitten off his tenth head, and his wound was still bleeding."

Other stories tell of a time when there were ten suns in the sky (10 signifying "a lot"), but nine of them were crushed between mountains and destroyed by the huntsman Yang Oerlang or, alternatively, shot down with arrows by the archer Hou I. That is why we have just one sun now.

Cats are very lucky creatures because they get nine lives. Or at least they do in the English-speaking world. Mexican, Brazilian, Spanish, and Iranian cats get seven lives, another auspicious number. The number 7, as well as being odd and prime, has an additional astronomical symbolism: before telescopes were invented, we could see seven astronomical bodies that, unlike the stars, seemed to move around freely in the sky. These were the sun, the moon, and the five closest planets: Mercury, Venus, Mars, Saturn, and Jupiter. The number 7 thus acquired an important significance. This, and the fact that four seven-day weeks fit nicely into one twenty-eight-day cycle of the moon, is almost certainly why we have a seven-day week, why many creation stories have the world being created in seven days, and probably also why, on a less exalted plane, Snow White meets seven dwarfs.

The number 5, unlike 7, has not an astronomical but an anatomical symbolism: it's literally a handful. The Five Pillars of Islam, the Five Symbols of Sikhism—these can be counted on the fingers of your hand. While in ancient Greek tradition there are four elements, in China there are five: fire, earth, metal, water, and wood. For geometers, 5, among the numbers that surround it, is an anomaly. An artist can make a regular tiling pattern with equilateral triangles, squares, or regular hexagons (bees can also do the last of these). But this cannot be done with regular pentagons. One thing you can do with five points, though, is make a star shape. Even better, you can draw it in a single continuous line, moving from point to opposite point, without taking your pencil off the paper. This is impossible for any smaller number, and if you try it for 6 you find the design splits into two triangles. Alchemical folklore has associated the five-pointed star known as the pentagram with such mischievous pursuits as summoning demons, because it was believed to be a protective talisman, preventing the demons from escaping its continuous boundary. In Goethe's play *Faust*, Mephistopheles cannot leave Faust's study because there is a pentagram drawn above the door. But hang on, asks Faust:

The pentagram prohibits thee?
Why, tell me now, thou Son of Hades,
If that prevents, how cam'st thou in to me?

Mephistopheles replies that the final line of the pentagram is incomplete—the outer angle was left unfinished, with the lines not quite meeting. Because of that small error, Mephistopheles has been able to materialize in the room, but enough of the pentagram has been drawn to prevent him going beyond it. The technique to construct a perfect pentagram with only a straightedge and compass has been known to mathematicians for at least two thousand years. If only Faust had been better at geometry, all that unpleasantness could have been avoided.

To finish the chapter, let's dive into all things three. The number 3 has an astonishing hold on Western minds. If you can get hold of it, I highly recommend the 1968 essay "The Number Three in American Culture" by the anthropologist Alan Dundes.[4] It lists a mind-boggling array of threenesses. In the world of nursery rhymes, trebling is common, either of words ("Row, row, row your boat") or phrases ("Do you know the muffin man, the muffin man, the muffin man"). This extends to common expressions—nobody gets "two cheers," after all. We learn our ABCs, not our ABCDs. Races start with "Ready, set, go," and three finishing positions are rewarded with gold, silver, and bronze. Three-letter abbreviations are everywhere: JFK, VIP, SOS, DNA, HBO, and let's not forget USA. Clothes come in small, medium, and large (or, if other sizes exist, they are given in reference to these three—XS, XXS, XL, XXL, and so on). Three-word phrases also abound: hook, line, and sinker; lock, stock, and barrel; wine, women, and song. We treble things for emphasis: the truth, the whole truth, and nothing but the truth. At the end of more than twenty pages, Dundes challenges the reader: "If anyone is skeptical about there being a three-pattern in American culture, let him give at least three good reasons why."

In terms of literature, the first aspect of threeness that we notice in narrative is sets of three characters. Three little pigs, three billy goats gruff, three good fairies, three bears. Countless tales involve three brothers being set a task. The first two try and fail; the third, the youngest, bravest, cleverest, and most underrated, succeeds. Or there are three sisters (as in "Beauty and the Beast"). The elder two are usually some charmingly misogynistic mix of vain, ugly, greedy, and stupid. Sometimes they are stepsisters, as in "Cinderella." The youngest, modest and beautiful, gets to marry the handsome prince. This pattern is also seen in jokes that take the form of a story involving three characters. Sometimes it's a minister, a priest, and a rabbi. Mathematicians sometimes make jokes against themselves that involve, for example, a physicist, an engineer, and a mathematician confronted with some problem.

In both the fairy tales and the jokes, the structure is that the same basic situation occurs twice with basically the same outcome, and then the third character encounters the situation and something different happens. In the joke, the two "normal" people react normally, then the fool does something ridiculous. In the fairy tale, this is inverted. The first two characters fail, and the third succeeds. The first two pigs, for example, build their houses of straw and twigs, but the third pig builds his house of brick. The first two brothers don't help the ugly old beggarwoman, but the youngest brother does. She inevitably turns out to be an enchantress in disguise, and she showers him with riches. The narrative reason for this is obvious—we require two repetitions to get to know the pattern so that the breaking of the pattern in the third iteration can surprise or amuse us.

Of course, it's not just in fairy tales that we encounter numbers. Dante's *Divine Comedy* has a huge amount of mathematical metaphor, with several numbers being accorded special significance. But it's the number 3 that is most fundamental to both its construction and its symbolism. No doubt the reason for this is the great spiritual importance, for Dante, of the Trinity (God the Father, God the Son, God the Holy Ghost). There are three books. *Purgatorio* and *Paradiso* have 33 cantos, while *Inferno* breaks the symmetry with 33 + 1 (well, it is hell, I suppose). This brings the total to 100. Each canto is a poem written in a style invented by Dante called *terza rima*: stanzas of three lines, with an interlocking rhyme scheme: *aba*, *bcb*, *cdc*, *ded*, *efe*, and so on, for as long as you like. (Each canto is rounded off by a single line rhyming with the middle line of the final three—here it would be *f*.) This interlocking scheme, as well as linking consecutive stanzas together in an elegant way, gives an additional threeness to the proceedings because apart from possibly the first and last rhymes, every other rhyme appears exactly three times. There are nine (three times three) circles of hell, split into three parts corresponding to the three main kinds of sin that get you admitted. Paradise also has nine circles, or nine heavens. And in the very last canto, Canto 33 of *Paradiso*, when Dante is on the verge of ascending to the vision of God, he sees "three circling spheres, three-coloured, one in span"—three perfect rainbows, in other words.

What can explain the hold that the number 3 has on our psyche? I propose that the mathematics of triangles and trichotomies enables the triumph of the triple. Three, in geometry, is very special. First, it's the smallest number of points that can define a two-dimensional shape. If you have only two points, then you get just a line. Three points (as long as they don't all lie on the same line) give you a triangle. But it's better than that. Imagine trying to make a rigid, stable structure out of sticks or rods. With two rods, you can't do anything. You can join two ends together, but the other two ends just flop about uselessly. But if I start with three rods of any lengths I like, I can fit them together in *exactly* one way to make a triangle. If you also have three rods of the same length and do the same thing, our two triangles will look the same. That's the second special thing about the number 3. It is not true for any higher number. With four rods, there are infinitely many quadrilaterals (four-sided shapes) you can make. Even in the superspecial case in which I want all four side-lengths to be the same, there are infinitely many possibilities. You can make a square, sure, but then you can squeeze it at the corners to make a series of ever-thinner diamonds. The triangle is the only straight-line shape that can't be deformed in this way. That's why in structures made with steel rods, such as geodesic domes, the basic shape is the triangle. It is the strongest shape.

The third (of course) special geometric property of the number 3 is that three is the largest number of points you can have in a plane that are all the same distance from one another. The three points of an equilateral triangle are mutually equidistant. It's impossible to draw four or more points on a piece of paper all of which are the same distance from each of the others. (You can do it for four points if you go to the third dimension, with what's called a *tetrahedron*, but even then, this is a shape made by joining four equilateral triangles together.) These geometric properties of the triangle could be one reason why sets of three things give us a sense of strength and completeness, and also often of equitability. All for one and one for all, as the Three Musketeers say. With two, it's just the up or down, left or right, or north or south of a line. With three, suddenly we can encompass a whole space.

The final mathematical aspect of 3 is the trichotomy. Imagine the whole number line laid out, and stick a pin in it at a point x. Every other number has a relation to x, and there are precisely three possibilities (a trichotomy). It is less than x, or it equals x, or it is greater than x. This kind of trichotomy is all over the place in mathematics. Every angle is either acute (less than 90 degrees), a right angle (equal to 90 degrees), or obtuse (greater than 90 degrees). Numbers are negative, positive, or zero. Time can be past, present, or future. In statistics, a data point can be higher than the mean, lower than the mean, or bang-on average.

A variant of this idea is the set of three we obtain from the two extremes plus the middle. Smallest, biggest, and everything in between. Sunrise, daytime, sunset. Birth, life, death. Trichotomies like this happen regularly, in both the language and the structure of narrative. We have three layers for adjectives: good, better, best; bad, worse, worst; brave, braver, bravest. The youngest of three fairy-tale brothers is invariably the wisest; the youngest sister is always the prettiest; the third billy goat gruff is the biggest and defeats the troll. And what better example of trichotomy than the threefold verdicts given by everyone's favorite housebreaker, Goldilocks? Daddy Bear's porridge is too hot. Mummy Bear's porridge is too cold. Baby Bear's porridge is just right. Goldilocks was clearly familiar with Aristotle's doctrine of the mean. He says that every ethical virtue is a golden mean (just right) between two vices—one an excess, the other a deficiency. Courage is a virtue, an excess of courage is the vice of recklessness, and a deficiency of courage is the vice of cowardice. When it comes to money, liberality is a virtue, an excess of liberality is profligacy, and a deficiency is miserliness. And when it comes to beds, Daddy Bear's bed is too hard, Mummy Bear's is too soft, and Baby Bear's epitomizes the Aristotelian mean—it's *just right.*

Stories themselves have a beginning, middle, and end. The most common multivolume set is the trilogy. These often are trilogies only in hindsight; a common structure is a self-contained initial volume, followed by a Book 2 that ends on a cliff-hanger, or at least with matters unresolved, and then a concluding Book 3 wrapping up all the loose ends, so that the trilogy is a larger-scale version of beginning, middle, end. Consider also the three-act play, in which each scene itself must also have a beginning, middle, and end. The book you are holding in your hands itself has three parts.

The appearance of magic numbers in fiction may be the most obvious manifestation of mathematics in literature, but that's just the beginning. Later, I'll show you the ways in which much more sophisticated mathematical ideas, from geometry to algebra and even calculus, have made their appearance in great works of literature, from *Moby-Dick* to *War and Peace*. Numbers are such a crucial part of human thought that they are even hidden inside words themselves, sometimes in the most unexpected places. Think of the fateful bowl of punch that dashes Becky Sharp's hopes for a proposal from Jos Sedley in *Vanity Fair*. No numbers there, right? Except that the word "punch" derives from the Sanskrit word for "five," *panca,* because the drink originated with an Indian concoction that had five ingredients. Numbers really are part of the fabric of language, in (to use the ancient Greek word for "ten thousand") myriad ways.

6

Ahab's Arithmetic

Mathematical Metaphors in Fiction

I mentioned in the introduction that the seeds for this book were planted when I heard a mathematician mention that *Moby-Dick* contains a reference to an interesting curve called a cycloid. Curiously enough, when I emailed my friend Tony (the mathematician in question) a couple of years back to thank him for his recommendation, he replied saying *I'd* been the one to recommend it to *him*—so I guess we'll never know the truth. At any rate, one morning I sat down on the train, opened the book, started reading, and within a few minutes encountered a brilliant description with a definite mathematical tinge to it. Ishmael spends the night at the Spouter-Inn, whose landlord is somewhat stingy with his drinks: "Abominable are the tumblers into which he pours his poison. Though true cylinders without—within, the villainous green goggling glasses deceitfully tapered downwards to a cheating bottom. Parallel meridians rudely pecked into the glass, surround these footpads' goblets." It's a great image, and there's an undeniably geometrical air to the true cylinder marked with parallel meridians. It piqued my interest.

As I read on, I kept encountering mathematical allusions, so many in fact that it became clear to me that Melville obviously relished mathematical ideas—they were bound to escape from his mind onto the page, and when he reached for a metaphor, more often than not something mathematical would present itself. In praising the loyalty of his cabin boy, Captain Ahab says, "True art thou, lad, as the circumference to its centre." And indeed this is exactly right—the points on the circumference of a circle steadfastly remain exactly the same distance from the center, all the way around.

The world of mathematics is a glorious source of metaphors. Some of these have become clichés in everyday speech, like "squaring the circle"—a reference to the ancient Greek problem of constructing a square with the same area as a given circle. Few people using this phrase know that the mathematical proof of its impossibility took more than two millennia to find. But sometimes you'll come across authors who, like Melville, clearly have an affinity for mathematics and cannot help but use mathematical metaphors in their writing. In this chapter, I'll give you a guided tour of some of the loveliest mathematical allusions in the work of classic writers like Melville, George Eliot, Leo Tolstoy, and James Joyce. Understanding these references adds another layer to our enjoyment of great literature, and it'll give you a totally new perspective on some much-loved books, as well as on their authors.

Before I show you any more of Melville's mathematical metaphors, I want to tell you a little about Melville himself and how he came to write what D. H. Lawrence described as "a surpassingly beautiful book . . . a great book, a very great book, the greatest book of the sea ever written. It moves awe in the soul." Melville tried various professions (teacher, engineer, deckhand on a whaling ship) before writing his first novel, *Typee*, a fictionalized account of his time with the Polynesian tribe of that name. This, and the follow-up, *Omoo* (the Polynesian word for "wanderer"), were well received, and he wrote three more seafaring stories over the next few years. I am focusing on *Moby-Dick*, his sixth book, because it's my favorite of Melville's books and it's the best known.[1] But Melville's love of mathematics seeps into everything he does. In his earlier novel *Mardi*, he has a character cry out, "Oh Man, Man, Man!

Thou art harder to solve, than the Integral Calculus." His publisher was clearly concerned that discussing philosophy and mathematics might not be as profitable as writing about scantily clad young Polynesian ladies, and Melville reassured him that the next book would contain "no metaphysics, no conic-sections, nothing but cakes & ale." Fortunately for literature, he comprehensively failed to keep that promise.

Moby-Dick was written in 1850 and published in 1851. Reviews were . . . mixed. *Harper's New Monthly Magazine* loved it: "The genius of the author for moral analysis is scarcely surpassed by his wizard power of description." But a reviewer for the London *Athenaeum* felt that "Mr. Melville has to thank himself only if his horrors and his heroics are flung aside by the general reader, as so much trash belonging to the worst school of Bedlam literature." It's amazing to think that the author of *Moby-Dick* more or less gave up writing within a few years of its publication; he spent the last two decades of his life working for the US customs service and died in obscurity in 1891. His lifetime earnings from perhaps the most influential American novel of the nineteenth century amounted to $556.37. We don't know much about Melville the man—as an indication of how carefully he guarded his privacy, he used to hang a towel over the doorknob of his study so that nobody could look through the keyhole. Few of his letters survive, and all that his close friend Nathaniel Hawthorne could find to say was that, although he was a gentleman, he was "a little heterodox in the matter of clean linen." But listen: if you've not done so already, please overlook the dirty laundry and read *Moby-Dick*. It's like no other book.

Our narrator, Ishmael, goes to work as a deckhand on a whaling ship, the *Pequod*, with its captain, Ahab; first mate, Starbuck (of coffee shop fame); and second mate, Stubb. Gradually it becomes clear that Ahab is obsessed with hunting and killing the great white whale Moby Dick—a previous encounter with whom led to Ahab's losing his leg. (By the way, the book is titled *Moby-Dick*, but the whale is named Moby Dick. If you are angry about the inconsistency, please take it up with Ishmael.) Ultimately, Ahab's hubristic and monomaniacal pursuit of the white whale drives him to insanity, endangering the whole crew, and let's just say it doesn't end well for Ahab.

This is no ordinary adventure story. There are "excerpts" discussing mentions of whales and whaling from a dizzying array of sources, including Shakespeare, the Bible, and books on natural history and navigation. There's a whole chapter on the meaning of Moby Dick's whiteness, and many philosophical musings from Ishmael and others. Ishmael explains that the book must necessarily have a huge compass because its subject, Leviathan, is so vast. "Give me a condor's quill!" he says. "Give me Vesuvius's crater for an inkstand!"

If I asked you to predict where mathematics might appear in a nineteenth-century sea story, you might think of quadrants and sextants and quite rightly suggest that it could be involved in descriptions of navigation. We do indeed hear about Ahab's doing mathematical calculations "on the upper part of his ivory leg," and Ishmael talks of "studying the mathematics aloft there" in the crow's nest where he perches, scanning the sea for signs of whales. But Melville goes a lot deeper. The almost magical powers of mathematics, for those initiates who can decipher its "cabalistic contrivances," are spoken of by the crew with a mixture of awe and suspicion: "I have heard devils can be raised with Daboll's arithmetic," says the second mate, Stubb. Generations of American schoolchildren would have been familiar with Daboll's *Arithmetic*, the most widely used textbook in US schools for the first half of the nineteenth century. (The full title is *Daboll's Schoolmaster's Assistant: Being a Plain, Practical System of Arithmetic, Adapted to the United States.*) The author, Nathan Daboll, was a Connecticut mathematics teacher, and we know that Melville used Daboll's *Arithmetic* as a pupil and probably as a teacher. You would have Daboll for arithmetic and Euclid for geometry.

Looking at the book with modern eyes, it's not at all surprising that Stubb compared it to some sort of alchemy. It provides techniques, which are to be learned by rote, for all manner of calculations, from the basics of arithmetic to currency conversion and rules for calculating interest, annuities, profit and loss, and ship tonnage. Methods are even given for the extraction of square roots and cube roots by hand. The rules given are often presented almost as magic formulae. For instance, to convert from South Carolina dollars to Maryland dollars, "multiply the given sum by 45, and divide the product by 28." Or there's the mysterious "Rule of three direct," which teaches, "By having three numbers given to find a fourth, which shall have the same proportion to the third, as the second has to the first." Here is the rule for finding the circumference of a circle, given its diameter: "As 7 is to 22, so is the given diameter to the circumference. Or, more exactly, as 115 is to 355; the diameter is found inversely." The circumference of a circle is its diameter d multiplied by π, but amazingly there is no mention of π here, or the fact that these rules work because $\frac{22}{7}$ and $\frac{355}{115}$ are approximations to π. They are just magic numbers to be deployed.

For Stubb, mathematics is mysterious, even malign. But for Ishmael, mathematics, and symmetry in particular, symbolize virtue. The sperm whale has a "pervading dignity" because of the "mathematical symmetry" of its head. In describing this head, Ishmael even claims to define a new mathematical concept. He explains, "Regarding the Sperm Whale's head as a solid oblong, you may, on an inclined plane, sideways divide it into two quoins, whereof the lower is the bony structure, forming the cranium and jaws, and the upper an unctuous mass wholly free from bones." In a footnote, he explains, "Quoin is not a Euclidean term. It belongs to the pure nautical mathematics. I know not that it has been defined before. A quoin is a solid which differs from a wedge in having its sharp end formed by the steep inclination of one side, instead of the mutual tapering of both sides." This could have come straight out of a geometry book!

You could argue that it's fair enough to get a bit geometrical when describing a shape (though it does indicate at least comfort and facility with these terms), but Euclid gets name-checked in several other places too. When explaining that the whale's eyes, being on opposite sides of its head, present its brain with two completely distinct views that must be processed simultaneously, Ishmael says that if the whale can really do this, then "it is as marvelous a thing in him, as if a man were able simultaneously to go through the demonstrations of two distinct problems in Euclid." The best mathematical moments in *Moby-Dick* are in places like this, where Melville throws in a mathematical allusion just for the fun of it.

It takes a geometer's eye, for instance, to connect a whale's fin to the gnomon of a sundial, as in Ishmael's observation here:

> *When the sea is moderately calm, and slightly marked with spherical ripples, and this gnomon-like fin stands up and casts shadows upon the wrinkled surface, it may well be supposed that the watery circle surrounding it somewhat resembles a dial, with its style and wavy hour-lines graved on it. On that Ahaz-dial the shadow often goes back.*

Pleasingly enough, that mention of Ahaz recalls what is now thought to be the earliest written reference to sundials, in the Old Testament book of Isaiah. God causes the shadow on a sundial to move miraculously backward ten degrees, as a sign that he will cure the sickness of Hezekiah, son of King Ahaz of Judah.

But perhaps the most fascinating bit of geometry in *Moby-Dick* involves cycloids, those mathematical curves I mentioned at the start of this chapter. Ishmael thinks about them while he is cleaning the great try-pots on the deck of the *Pequod.* Try-pots are huge metal vats—think massive cauldrons—where the whale blubber is rendered to produce oil:

> *Sometimes they are polished with soapstone and sand, till they shine within like silver punchbowls. . . . While employed in polishing them—one man in each pot, side by side—many confidential communications are carried on, over the iron lips. It is a place also for profound mathematical meditation. It was in the left hand try-pot of the* Pequod, *with the soapstone diligently circling round me, that I was first indirectly struck by the remarkable fact, that in geometry*

all bodies gliding along the cycloid, my soapstone for example, will descend from any point in precisely the same time.

Inverted cycloid—an object released from any point will take the same time to slide to the bottom

A cycloid, if you recall, is the curve traced out by a point on the edge of a rolling circle or wheel:

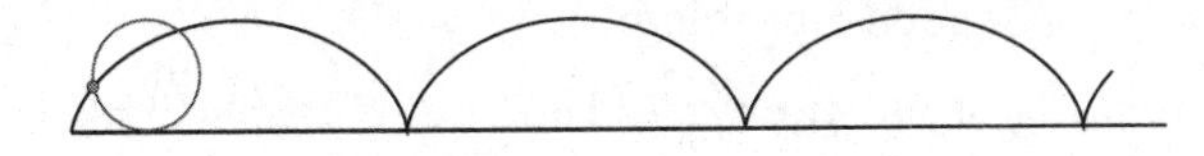

It's not something that has ever been routinely taught in schools, but it's one of the most famous curves in mathematics. I said in the introduction that it was nicknamed "the Helen of geometry" because of its beautiful properties, but that's only part of the story. The epithet also alludes to the many squabbles it caused among competing mathematicians. The list of people who studied it reads like a roll call of seventeenth-century mathematicians, including René Descartes, Isaac Newton, and Blaise Pascal. Pascal, a brilliant mathematician, more or less invented the mathematical study of probability.[2] At one point he stopped studying mathematics in favor of theology. But one night he had a terrible toothache. To distract himself, he started thinking about cycloids, as one does, and to his surprise the pain went away. Naturally, he took this as a sign that God was okay with his interest in mathematics, and he kept thinking about cycloids for eight more days, during which time he discovered many of their properties—things like the area under their arches.

Mathematicians in the past frequently argued about who had proved things first, and priority disputes could get heated. (Such things are easier to resolve now, with an electronic trail of breadcrumbs to follow.) A mathematician named Gilles de Roberval, for example, proved lots of things about the cycloid, but he refused to publish any of them. Then, whenever someone announced a new result, Roberval would angrily retort that he had established that ages ago. Roberval knew, for example, that the area under any arch of the cycloid is exactly three times the area of the circle that made it.

Part of the reason for this silly behavior was that Roberval's job was a professorship that was reappointed every three years via a competition whose problems were set by the incumbent professor. So there was a strong incentive to have a set of problems that only you knew how to solve. Finding the area under a given cycloid would have been, at least for a few years, just such a problem.

The best thing about the cycloid for me is that it makes an unexpected appearance in a context that seems to have nothing to do with the way it's constructed. In trying to improve clock design, the Dutch mathematician Christiaan Huygens was interested in finding out whether there is a curve with the property that something sliding down the curve would get to the bottom of the curve in exactly the same time no matter where it started from. This is called the *tautochrone problem*, and he solved it in 1659—you can read all about it in his 1673 blockbuster *Horologium Oscillatorium*. There's another problem, called the *brachistochrone problem*: given two points, find the path between them that would get something from the higher point to the lower point, falling under gravity, in the quickest time. Amazingly, the solution to both these problems is our friend the cycloid!

It's the tautochrone problem that Ishmael is talking about. If you have a try-pot in the shape of one of those cycloid arches (turned upside down, of course), then wherever you release your soapstone from, it will take exactly the same amount of time to reach the bottom. Specifically, the time of descent is always $\pi\sqrt{\frac{r}{g}}$ seconds (that g is acceleration due to gravity, and r is the radius of the circle that made the cycloid). What's brilliant is that, on Earth, g is about 9.8, and the square root of that is about 3.13. Since that's on the bottom of the fraction, and π, which is about 3.14, is on the top, these almost exactly cancel out. This means that, to quite a good approximation, the time of descent in the cycloid is just the square root of the radius of the circle it's made from. Wow!

How did Melville/Ishmael know this? We can't be sure—it wouldn't have been in the standard school curriculum of the time. But a researcher named Meredith Farmer discovered that there was a rather exceptional mathematics teacher at the school young Herman attended in 1830 and 1831, the Albany Academy. Records state that every afternoon there was spent on "arithmetic": each student "is employed one hour in his arithmetic lesson and engaged during the remainder of the Afternoon in entering sums into a large ciphering book." This doesn't sound too promising. But Farmer realized that Herman's teacher for these classes was none other than Joseph Henry, professor of mathematics and natural philosophy (as natural science was then known), a brilliant teacher and at the same time a well-known scientist who would go on to be the first secretary of the Smithsonian. He discovered inductance, which is why the unit of inductance is called the henry. Herman excelled in these lessons and won a prize for being "the first best in his class in ciphering books" (his prize was a book of poetry). Joseph Henry in fact wrote to the academy board a few months before Herman won his award, asking to add a more advanced textbook for the "higher students." He was a passionate, inspiring teacher—some of his more advanced lessons were even given as public lectures. I can't prove it, but there's every chance Henry taught Melville and the other "higher students" about cycloids and nurtured Melville's enthusiasm for mathematics.

There's a broader mathematical theme underlying *Moby-Dick*, and that's the symbolism of mathematics as a way of understanding, and to some extent trying to control, our environment. Mathematics helps us to navigate the unknowable universe. Ishmael certainly values data: his own body is used to record them. He tells us that "the skeleton dimensions [of the whale] I shall now proceed to set down are copied verbatim from my right arm, where I had them tattooed; as in my wild wanderings at that period there was no other secure way of preserving such valuable statistics." But it is a mistake to assume that analysis is the same as control, just as it is a mistake to reject mathematics completely. Ahab veers between the two extremes. He studies charts and records of whale sightings obsessively, convinced that he can predict where Moby Dick will turn up. But later, as his madness grows, he rejects the mathematical calculations of navigation, trampling his quadrant into pieces and ultimately sailing on instinct alone. Mathematics is abandoned, leaving us adrift in the ocean.

Ahab's obsession with Moby Dick leads him to the irrational belief that by knowing the behavioral patterns of whales in general, he can somehow derive certain knowledge about a specific whale. This seems a dubious enough proposition with regard to whales, but the patterns of human society are even more complicated. To what extent can information about the population as a whole tell us anything about a single person? The interplay between the actions of individuals and the broader statistical sweep of events is a major theme for another nineteenth-century novelist, George Eliot. Her 1876 novel, *Daniel Deronda*, begins in a casino, with Gwendolen Harleth playing roulette, the outcome of which is governed by the laws of probability. Life outcomes, however, cannot be predicted. If we trust that the next throw of the dice will go in our favor, we are likely to be disappointed. In *Silas Marner* (1861), another Eliot novel full of chance and randomness, the community trusts that drawing lots to determine whether Silas Marner is guilty of theft will result in a true verdict. But that's not what happens.

Chance events, at the gambling table or in life, have certain probabilities of occurring, but even so, there is no way of knowing to whom they will happen and when. Statistics was in its infancy as a mathematical science in the nineteenth century. The word "statistics" comes from the German word *Statistik*, meaning something like "science of the state." In English, it used to be called "political arithmetic." Originally, it was not much more than counting: What is our population? What is our production of wheat each year? Statistical analysis came later, importing the techniques from probability to explore chance, and the types of data that were considered were broadened significantly to include things like crime statistics or causes of death. This led to a lot of soul-searching about the implications for free will and fate. Charles Dickens was troubled by the law of averages. If the number of people killed so far this year is below the annual average, he wrote, "is it not dreadful to think that before the last day of the year some forty or fifty persons must be killed—and killed they will be." Even something as deeply personal as the decision to end your own life, says the French sociologist Émile Durkheim in his 1897 book *Suicide: A Study in Sociology*, is nevertheless part of a "collective tendency."

There's a scene in *Daniel Deronda* in which Daniel and his friend Mordecai join a discussion in a pub:

> *But to-night our friend Pash, there, brought up the law of progress, and we got on statistics; then Lilly, there, saying we knew well enough before counting that in the same state of society the same sort of things would happen, and it was no more wonder that quantities should remain the same than that qualities should remain the same, for in relation to society numbers are qualities—the number of drunkards is a quality in society—the numbers are an index to the qualities, and give us no instruction, only setting us to consider the causes of difference between social states.*

Numbers are "qualities" (that is, properties)—numbers have the power to tell us about society. But they still cannot tell us the fate of an individual, just as knowledge of the probabilities of roulette cannot tell us whether the next number will be red or black.

Daniel Deronda, with its themes of gambling and fate, explores this idea on both the micro and macro levels. Gwendolen, who loses her money at the gambling table that night, comes home to the news that her family's fortune has been lost in the vicissitudes of the economy. Her decision to marry the dreadful Grandcourt is described as a roulette-like gamble too. The variations in Gwendolen's fortunes through the novel are spins of the wheel on a larger scale. She wins at roulette, then loses. Her family is rich, then ruined. She enters into an unhappy marriage, which ends with Grandcourt's accidental death. But even though ideas of probability might suggest that the bad (losing, being ruined, an unhappy marriage) would be balanced out in the long term by the good (winning, being wealthy, a happy marriage), this novel gives the lie to the idea that balance must be restored within any particular time period. At the end of the novel, Gwendolen does not get to marry Daniel, the man she loves. We do not find out the rest of her story. All we have is her resolution that "I shall live. I shall be better."

George Eliot (her real name was Mary Ann Evans) was born in 1819, the same year as Herman Melville. She had an abiding interest in mathematics, and though unlike Melville she didn't have the opportunity for a formal schooling in mathematics, it's clear from both her novels and her surviving letters and notebooks that her knowledge of the subject was considerable.[3] She constantly reaches for mathematics to illuminate her thoughts. In fact, a student at the University of Leicester, Derek Ball, wrote his entire Ph.D. dissertation on the mathematics in George Eliot's novels.[4] Once you are on the lookout, you'll see it everywhere. In *Middlemarch* (1871–72), the inconsistency of Mr. Brooke's largesse is satirized mathematically: "We all know the wag's definition of a philanthropist: a man whose charity increases directly as the square of the distance." When Mr. Brooke is trying to work out why on earth his lovely young niece Dorothea would want to marry boring old Edward Casaubon, he concludes that "woman was a problem which . . . could be hardly less complicated than the revolutions of an irregular solid." Daniel Deronda himself studies mathematics at Cambridge—"the study of the higher mathematics, having the growing fascination inherent in all thinking which demands intensity, was making him a more exclusive worker than he had been before."

The mathematical knowledge that Eliot displays goes far beyond the superficial. Take, for instance, the introduction of Mr. Casson, landlord of the Donnithorne Arms, in her first novel, *Adam Bede* (1859):

> *Mr. Casson's person was by no means of that common type which can be allowed to pass without description. On a front view it appeared to consist principally of two spheres, bearing the same relation to each other as the earth and the moon: that is to say, the lower sphere might be said, at a rough guess, to be thirteen times larger than the upper. . . . But here the resemblance ceased, for Mr. Casson's head was not at all a melancholy-looking satellite, nor was it a "spotty globe," as Milton has irreverently called the moon.*

It's a pleasing metaphor, and we can instantly form a mental picture of Mr. Casson from it. But as Derek Ball points out, the specific use of the number 13 betrays a significant mathematical literacy. First, we have to ask to what it refers. The diameter of the earth is 7,918 miles, which is about 3.7 times the 2,159-mile diameter of the moon, so it's not that. The volume of the earth is about 49 times that of the moon, so it's not that either. But when we look "on a front view," as Eliot says, at two spheres, what we see is in fact two circles. So what our brains perceive intuitively in this situation is likely to be the respective sizes, or areas, of these circles. Lo and behold, the cross-sectional area of the earth is 13.45 times that of the moon, hence the number 13. What's even more impressive is that Eliot must have used a good approximation for the diameters. If you try this calculation with a starting point of 8,000 miles for the diameter of the earth and 2,000 for the moon, you'd get a ratio of 16 times, not 13. Even if you use my "about 3.7 times" approximation, you arrive at a number closer to 14 than 13. What does this tell us? That Eliot chose a mathematical image, that she made a sensible choice about which ratio to choose, and that she was able to calculate that ratio to a high level of accuracy.

George Eliot's interest in mathematics was lifelong. She had a large acquaintance among the scientists and mathematicians of the day, and she kept extensive notebooks full of interesting observations on a wide variety of topics. Many, though, are mathematical. On probability, for example, she recounted a curious phenomenon known today as *Buffon's needle*. Imagine you have wooden floorboards. If you drop a needle onto the floor, and if the needle's length equals the width of the boards, then the probability that the needle will lie across two floorboards is exactly $\frac{2}{\pi}$. You can actually use this expression to find an approximation for π by performing the experiment repeatedly. If you drop the needle 25 times and it falls across two floorboards 16 times, then you have obtained the approximation $\frac{16}{25}$ for $\frac{2}{\pi}$, which gives that $\pi \approx 3.13$. Not far off.

Eliot studied mathematics both informally and formally, including attending a course of twice-weekly geometry lectures in 1851. Even in the final year of her life, she was still actively learning mathematics, telling a friend that she was studying conic sections every morning. (Conic sections are the different curves that can be made by slicing through a cone: the parabola, ellipse, and hyperbola. I've always found it kind of cute that each of these curves also gives us an adjective to describe writing, which can be parabolic, elliptical, or hyperbolic.) Her novels reflect her interest in contemporary mathematical and scientific discoveries. In fact, Henry James actually criticized *Middlemarch* for being "too often an echo of Messrs. Darwin and Huxley." But it was to mathematics that Eliot would turn for consolation, especially at times of stress. A letter of hers from 1849 recounts her recipe for recovering from a difficult period in her personal life: "I take walks, play on the piano, read Voltaire, talk to my friends, and just take a dose of mathematics every day."

The habit of taking solace in the reassuring certainty, the eternal truth, of mathematics is something that is shared by Adam Bede in

Eliot's novel of the same name. When Adam's father dies and Adam is telling himself that life must go on, it's to mathematics that he reaches for a comparison. As he says, "The square o' four is sixteen, and you must lengthen your lever in proportion to your weight, is as true when a man's miserable as when he's happy."

The idea that mathematics can be a soothing balm for the tribulations of life can also be found in the work of a writer immeasurably far removed from Eliot's world: the Russian author Vasily Grossman. His 1959 masterwork, *Life and Fate*, was described in a 2021 *New York Times* essay by the editor and writer Robert Gottlieb as "the most impressive novel written since World War II." This epic novel recounts the story of brilliant physicist Viktor Shtrum and his family against a backdrop of war, the Battle of Stalingrad, and Communism. Grossman studied mathematics and physics at university, and the name Shtrum was chosen as a tribute to his friend, the real-life physicist Lev Yakovlevich Shtrum (1890–1936), who was executed during Stalin's "great purge." In the novel, Shtrum finds mathematics and equations a lodestone of rationality to cling to in a chaotic world:

> *His head had been full of mathematical relationships, differential equations, the laws of higher algebra, number and probability theory. These mathematical relationships had an existence of their own in some void quite outside the world of atomic nuclei, stars, and electromagnetic or gravitational fields, outside space and time, outside the history of man and the geological history of the earth. . . . It was not mathematics that reflected the world; the world itself was a projection of differential equations, a reflection of mathematics.*

With this outlook, it is mathematics that is the true reality, and everything else is simply a pale imitation. It's not possible in real life to construct an absolutely perfect circle, for example, but the mathematical concept of a circle encapsulates a higher truth. This higher truth is what is real to Shtrum, and its very perfection soothes the soul.

. . .

Life is messy. History is messy. Humans will keep acting unpredictably. For both Viktor Shtrum and George Eliot, mathematics is an escape from all that. But for Leo Tolstoy in *War and Peace*, mathematics is a way to coerce the chaos into sense. He uses mathematical allusions several times in the novel, but unless I want my book to be as long as his, I must restrict myself to mentioning just two.

When Stephen Hawking was writing *A Brief History of Time*, he was famously warned that every equation he included would cut sales in half. Well, he got away with it (I hope I do too), and so did Tolstoy when he invented an equation and dropped it into the middle of the action in *War and Peace*. Let me tell you about it.

When the French retreat from Moscow, they keep coming off worse in skirmishes with small groups of Russian troops, even though the French army is huge. This, says Tolstoy, seems to contradict conventional military wisdom that believes the strength of an army depends on its size alone. That, he says, is like claiming that momentum depends only on mass, when in fact it is the product of mass and velocity. In the same way, the strength of an army must be the product of its mass and some unknown x. Military science usually puts this unknown factor down to the genius of the commanders. But, says Tolstoy, the sweep of history is not decided by individuals. This x is rather the "spirit of the army, that is to say, the greater or lesser readiness to fight and face danger felt by all the men composing an army, quite independently of whether they are, or are not, fighting under the command of a genius." Like any good math teacher, Tolstoy even gives us an example. Suppose ten men (or battalions, or divisions) defeat fifteen, sustaining four casualties. Then the winning side loses four to the fifteen lost by their opponents "and therefore $4x = 15y$. Consequently $\frac{x}{y} = \frac{15}{4}$." This equation, as Tolstoy rightly points out, does not tell us what x and y are, but it does give us a ratio between them. Because $\frac{15}{4} = 3.75$, we can now say that the winning army has 3.75 times as much fighting spirit as the losing side. And, he concludes, "by bringing variously selected historic units (battles, campaigns, periods of war) into such equations, a series of numbers could be obtained in which certain laws should exist and might be discovered." Ah, yes, the classic final sentence of every application ever made for renewal of grant funding: "More research is needed."

If it's a surprise to see Tolstoy detonate an equation in the middle of the Napoleonic battlefield, then wait until he brings the big guns out: he uses calculus as a metaphor for understanding the whole of human history. In *War and Peace*, he argues powerfully against the idea that the course of history can be altered by the actions of any one person. The French army, he says, does not retreat from Moscow toward Smolensk because Napoleon has ordered it. Rather, Napoleon gave the order to retreat because "the forces which influenced the whole army and directed it along the [Smolensk] road acted simultaneously on him also."

So how do we make sense of these historical forces? Tolstoy begins by reminding us of that old puzzle of Achilles and the tortoise—known as *Zeno's paradox*. Achilles runs ten times as fast as the tortoise, so he should win any race, even if he gives the tortoise a head start. But in the time it takes Achilles to catch up to where the tortoise started, the tortoise has moved a little farther forward. And by the time Achilles reaches that next place, the tortoise has moved again. It seems that Achilles can never overtake the tortoise—which is obviously ridiculous. The paradox, says Tolstoy, is caused by the fact that the movement of Achilles and the tortoise is being artificially divided into discrete, discontinuous parts, whereas in reality, the motion of both is continuous. Fortunately, there is a branch of mathematics that tells you exactly how to turn the discrete into the continuous.

Calculus was developed in the late seventeenth century by two of the all-time mathematical greats, Isaac Newton and Gottfried Leibniz. (There were bitter arguments about who had thought of it first.) It's fantastic for solving problems that involve movement and change, like the motion of the planets, or objects accelerating under gravity (which is the other thing Newton is famous for). If something is moving at a fixed speed, we can work out how far it will travel—if it's going 40 miles per hour, then after an hour, it has gone, well, 40 miles. But what if the speed is constantly changing? How can we work it out then? What we could try is measuring the speed every minute, say, and then assuming that's the speed for the whole minute, working out the distance traveled in that minute, and adding up all those little distances. If we want to get more accurate, we could measure the speed every 30 seconds, or every second, or every nanosecond. Each time we are adding up ever tinier distances, summing over an ever larger number of these tiny changes, or "differentials." But the challenge you face is that in the limit, you'd be trying to add an infinite number of zeros. Calculus is the technique that allows us to deal with these infinitesimal numbers rather than imposing an artificial division into separate units. It is one of the great achievements of mathematics.

Tolstoy explains that we need to do the same thing with history: "The movement of humanity, arising as it does from innumerable arbitrary human wills, is continuous. To understand the laws of this continuous movement is the aim of history. But to arrive at these laws, resulting from the sum of all those human wills, man's mind postulates arbitrary and disconnected units," such as particular events in isolation, or the actions of some king or commander. "Only by taking infinitesimally small units for observation (the differential of history, that is, the individual tendencies of men) and attaining to the art of integrating them (that is, finding the sum of these infinitesimals) can we hope to arrive at the laws of history."

In *War and Peace*, Tolstoy is railing against the "great man" theory of history. It's not the great leader who is the *x* factor for the victorious army, but the collective fighting spirit. It's not the king or emperor who directs the course of events, but wider forces. He counteracts the theory with his fighting spirit equation and with the calculus metaphor we've just discussed. For him, mathematics is an emblem of logical rigor, a way to access the objective truth, and the only chance we have of understanding history.

. . .

War and Peace, with its mixture of history, philosophy, and narrative, was unlike any other novel. Tolstoy, in fact, said he didn't think of it as a novel at all. I want to finish this chapter with a look at the mathematics in another uncategorizable book: James Joyce's *Ulysses*.

I mentioned in the introductory chapter that Joyce had an admiration for mathematics, but if we think about what he is famous for, the stream-of-consciousness style of books like *Ulysses* and especially *Finnegans Wake*, he might be the author we would least think of associating with structure of any kind, never mind mathematics. And yet there's a diagram from Euclid right in the center of *Finnegans Wake*. There's a chapter full of calculations in *Ulysses*.

Geometry gets a mention in the first paragraph of the first page of Joyce's first published book, *Dubliners*: "Every night as I gazed up at the window I said softly to myself the word paralysis. It had always sounded strangely in my ears, like the word gnomon in the Euclid and the word simony in the Catechism." This mention of the gnomon is not a random allusion, either. The word is mostly used nowadays, if it's used at all, to refer to the sticking-up part of the sundial that casts a shadow (as we saw in *Moby-Dick*), but its geometric meaning is a parallelogram with a smaller parallelogram cut out of it. This "shape with a missing part" is a good description of *Dubliners*. Sometimes the missing part in a story is around meaning—the language used is ambiguous and we cannot see the motivations of the characters. Other times it is parts of the action that are omitted. In one story we are with a young woman, Eveline, at home, until she stands up suddenly, and the narrative jumps to a scene somewhere entirely different. We aren't party to her decision to leave the house, or where to go, or how she gets there.

Joyce had a reverence for, even an awe of, mathematics. Like Melville, he studied Euclid's geometry at school. Though he was not such a star student as Melville, Joyce was certainly familiar with algebra and geometry, and his extensive notebooks reveal a fascination with mathematical ideas.

He was curious about concepts of limits and infinity—there's a page on which he writes things like $0 = \frac{1}{\text{many}}$, $1 = \frac{1}{1}$, $\infty = \frac{\text{many}}{1}$. These represent limits, because if we divide 1 by ever larger numbers, we approach, but do not ever quite reach, zero, and the same is true for the characterization of infinity as $\frac{\text{many}}{1}$. Sometimes there is pseudomathematics too, as in the jotting $\text{JC} = \sqrt[3]{\text{God}}$, which is a rather facetious "formula" about the Holy Trinity.

Writers on Joyce have sometimes used mathematical analogies to describe his work, but not perhaps for the reasons I might do so. The writer of this 1941 obituary, for example, doesn't seem to have a very clear idea of what mathematics actually is:

> *Joyce was also the great research scientist of letters, handling words with the same freedom and originality that Einstein handles mathematical symbols. The sounds, patterns, roots and connotations of words interested him much more than their definite meanings. One might say that he invented a non-Euclidean geometry of language; and that he worked over it with doggedness and devotion.*

I have some issues with this. Firstly, Einstein didn't just say "Ooh, I think an m next to this c^2 would look cool." It's not the handling of the symbols that Einstein was good at, it's the meaning of the concepts. It reminds me of the time I was asked to prettify an equation for a newspaper article. Apparently, the graphic design department said it didn't look exciting enough visually—could I zhuzh it up a bit? I told them, not if you want it still to be true. Secondly, what could a "non-Euclidean geometry of language" possibly be? The obituarist simply grabbed a clever-sounding term from mathematics to say that Joyce did something exciting and new.

In this century, non-Euclidean geometry, while still very exciting, is not new. Nowadays, we get told that Joyce invented fractals. (We'll delve more into fractals in Part III.) I read an essay recently that posits the fractal (a new and exciting math concept circa 1980–2000) as "an active Joycean concept" and credits Joyce with "anticipating a fractal formalism that would not be officially discovered until well into the latter half of the twentieth century." For me, this goes too far. We have to be very careful about crediting writers with this kind of fortune-telling. Let me give an over-the-top example to prove a point. The scientist Murray Gell-Mann described how James Joyce provided the name for a new kind of subatomic particle discovered in the 1960s: "In one of my occasional perusals of *Finnegans Wake*, by James Joyce, I came across the word 'quark' in the phrase 'Three quarks for Muster Mark.' . . . The number three fitted perfectly the way quarks occur in nature." (For example, every proton contains three quarks.) Do we conclude from this that Joyce anticipated quantum physics? Of course not—and we shouldn't go around saying he anticipated fractals either. It's a shame, because as an analogy for what Joyce does in *Ulysses*, fractals are great. Zoom in as far as you like into the human experience, one might say, and the complexity is not diminished. The mind's experience of a single day, a single hour, is as richly detailed as the memories of a lifetime. Notwithstanding this fact, James Joyce didn't invent fractals. He doesn't need to have done that to be brilliant.

So, what can a conversation between James Joyce and mathematics tell us? Is it just that the work of Joyce is so dense with both meaning and ambiguity that we can put any meaning we like into it? In the case for the defense, I bring to your attention Joyce's own words: one entire chapter of *Ulysses* was, he said, a "mathematical catechism." I want to explain that a little.

You might remember that *Ulysses* is loosely based on Homer's *Odyssey*, an epic poem recounting the adventures of Odysseus, king of Ithaca, over ten years as he travels home after the Trojan Wars. The name Ulysses is the Latinized version of Odysseus. The action in Joyce's book is transplanted to Dublin and describes the events of one fairly ordinary day in the life of one fairly ordinary middle-aged man, Leopold Bloom (Ulysses), a young man he meets, Stephen Dedalus (representing Telemachus—Odysseus's son), and Bloom's wife, Molly (Penelope). Each chapter is associated in some way with an episode from the *Odyssey*: Chapter 11 is known as "Sirens" and is full of singing and music; Chapter 17 is known as "Ithaca" because it describes Bloom returning home at the end of the day, accompanied by Stephen Dedalus; and the final chapter in the book is "Penelope," with Molly Bloom's famous stream-of-consciousness monologue as she falls asleep.

What does mathematics do for James Joyce in *Ulysses*? There are mathematical references scattered throughout the book, but "Ithaca" is the most overtly mathematical chapter. It is, says Joyce, a "mathematico-astronomico-physico-mechanico-geometrico-chemico sublimation of Bloom and Stephen . . . to prepare for the final amplitudinously curvilinear episode *Penelope*." He goes further: it is best read by "someone who is a physicist, mathematician and astronomer and a number of other things." The structure of "Ithaca" is a series of questions—a catechism—that parodies scientific certainty. The books of Euclid were a cornerstone of mathematical education in Jesuit schools, and they were held up for millennia as the apotheosis of pure logic. The joke in "Ithaca" is the attempt to apply this logic to things that definitely aren't behaving rationally.

Stephen Dedalus and Leopold Bloom's nocturnal wanderings around Dublin are given a pseudogeometric veneer of respectability here in the opening question and response:

> *What parallel courses did Bloom and Stephen follow returning?*
>
> *Starting united both at normal walking pace from Beresford place they followed in the order named Lower and Middle Gardiner streets and Mountjoy square, west . . . they [crossed] the circus before George's church diametrically, the chord in any circle being less than the arc which it subtends.*

In other words, they took a shortcut across the circle, as it's quicker than going around. When they arrive home, it is to "the 4th of the equidifferent uneven numbers," which is Joyce's way of saying that

Bloom's house number is seven. Bloom lights a fire using "irregular polygons" of coal. In the kitchen, there are "four square handkerchiefs folded unattached consecutively in adjacent rectangles," suspended from a "curvilinear rope." It reads like a crazy math problem. Joyce really goes to town with all this a few pages later. Stephen is younger than Bloom, and the disembodied questioner would like to know "what relation existed between their ages." The answer to this is glorious:

> *16 years before in 1888 when Bloom was of Stephen's present age Stephen was 6. 16 years after in 1920 when Stephen would be of Bloom's present age Bloom would be 54. In 1936 when Bloom would be 70 and Stephen 54 their ages initially in the ratio of 16 to 0 would be as 17½ to 13½, the proportion increasing and the disparity diminishing according as arbitrary future years were added, for if the proportion existing in 1883 had continued immutable, conceiving that to be possible, till then 1904 when Stephen was 22 Bloom would be 374 and in 1920 when Stephen would be 38, as Bloom then was, Bloom would be 646 while in 1952 when Stephen would have attained the maximum postdiluvian age of 70 Bloom, being 1190 years alive having been born in the year 714, would have surpassed by 221 years the maximum antediluvian age, that of Methuselah, 969 years, while, if Stephen would continue to live until he would attain that age in the year 3072 A.D., Bloom would have been obliged to have been alive 83,300 years, having been obliged to have been born in the year 81,396 B.C.*

It all reminds me of a mathematical puzzle posed by Gustave Flaubert (an author much admired by Joyce) in an 1841 letter to his sister Caroline: "Since you are now studying geometry and trigonometry, I will give you a problem. A ship sails the ocean. It left Boston with a cargo of wool. It grosses 200 tons. It is bound for Le Havre. The mainmast is broken, the cabin boy is on deck, there are 12 passengers aboard, the wind is blowing East-North-East, the clock points to a quarter past three in the afternoon. It is the month of May. How old is the captain?" You are getting a lot of information here, but none of it actually helps you to solve the problem. We are back to the data overdose of Captain Ahab.

The stream-of-consciousness style of much of *Ulysses*, and orders of magnitude more in *Finnegans Wake*, belies the fact that every word is nonetheless carefully chosen.[5] Bloom's inner monologue through the day is full, like everyone's, of half-facts, snatches of quotations, fragments of misremembered science. The "Ithaca" chapter positions itself as authoritative, but Joyce inserts a huge number of errors that get in under the radar of the catechistic style. It reminds us that even sources like dictionaries and encyclopedias are not infallible. They are, after all, written by people. (My favorite dictionary definition of all time, by the way, is in the British *Chambers Dictionary* I have on my shelf, which defines an éclair as "a cake, long in shape but short in duration.")

Like the scientific "facts" in "Ithaca," many of the numerical calculations are incorrect. Some are incorrect on purpose, some probably aren't. When Leopold Bloom sits down at the end of the day and tallies up his expenditure, the fact that he "forgets" to write down the money spent in a brothel is not an error in Joyce's arithmetic. But there are also several miscalculations around Bloom's and Stephen's ages and the year Bloom would have had to be born to achieve the correct proportions. For example, for Bloom to be 1,190 years old (seventeen times Stephen's age of seventy) in 1952, he would have been born in the year 762, not 714. We can see where the mistake comes from—if Bloom was born in the year 714, he would reach the age of 1,190 in 1904, when the book is set. But that would not preserve the 17:1 ratio of their ages. Even if these are deliberate mistakes, the number of corrections Joyce made to the calculation of Bloom's budget over the course of several drafts and proofs of the novel is good evidence that he did have some difficulty in manipulating the numbers, in spite of having performed relatively well in arithmetic exams at school.

But arithmetic is not mathematics, just as spelling is not literature, and there is a lot more than just arithmetic in "Ithaca." I want to show you a fun digression about powers because it resulted in a certain kind of number being named after James Joyce. Here is Leopold Bloom, thinking about the numbers involved in calculations about distances between the stars:

> *Some years previously in 1886 when occupied with the problem of the quadrature of the circle he had learned of the existence of a number computed to a relative degree of accuracy to be of such magnitude and of so many places, e.g., the 9th power of the 9th power of 9, that, the result having been obtained, 33 closely printed volumes of 1,000 pages each of innumerable quires and reams of India paper would have to be requisitioned in order to contain the complete tale of its printed integers of units, tens, hundreds, thousands, tens of thousands, hundreds of thousands, millions, tens of millions, hundreds of millions, billions, the nucleus of the nebula of every digit of every series containing succinctly the potentiality of being raised to the utmost kinetic elaboration of any power of any of its powers.*

Now, this is slightly silly of Bloom, because he doesn't actually have to do the calculation to know that the 9th power of 9 (or 9^9), whatever it is (okay, it's 387,420,489), is definitely less than the 9th power of 10, which is 1,000,000,000. So, the 9th power of the 9th power of 9 is going to be less than $(1{,}000{,}000{,}000)^9$, which is a 1 with 81 zeros after it. (It's 196,627,050,475,552,913,618,075,908,526,912,116,283,103, 450,944,214,766,927,315,415,537,966,391,196,809, if you want to know.) But let's be kind and assume that what Bloom meant to say was not the 9th power of the 9th power of 9, but 9 to the 9th power of 9. It's a curious fact about powers that if you try to do a power of a power, you have to be really careful what you mean by it. What is 3^{3^3}? Does it mean 3^3, which is 27, raised to the power 3? This would be $27 \times 27 \times 27$, which is 19,683. Or does it mean 3 raised to the power 3^3, which is 3^{27}, or just over 7.5 trillion? With exponents, it really matters where you put your brackets: $(3^3)^3 \neq 3^{(3^3)}$.

In honor of Joyce, mathematicians have named numbers like $3^{(3^3)}$ Joyce numbers. The *n*th Joyce number is $n^{(n^n)}$. If you thought powers of two grew quickly, these Joyce numbers, being exponents of exponents, grow even faster. The first Joyce number, $1^{(1^1)}$, is 1. The second, $2^{(2^2)}$, is 16. The third is 7.5 trillion, and the fourth is already too long sensibly to write, with 155 digits. If Bloom had been thinking of the ninth Joyce number, $9^{(9^9)}$, then he wasn't too far off with his estimate of the number of books required to contain it. It's possible that Joyce had read about this number somewhere because in 1906 the mathematician C. A. Laisant proved that $9^{(9^9)}$ has 369,693,100 digits. Over the thirty-three volumes of a thousand pages that Bloom recalls, this would mean squeezing around eleven thousand digits onto each page—just about possible, with a tiny typeface, no line spacing, large pages, and narrow margins.

This is certainly not the only large number in the Joycean oeuvre, and it is a mathematically sophisticated example of the tradition of those "upper limit" numbers like 99 and 999 that we saw in Chapter 5, because the number $9^{(9^9)}$ is very large but not infinite. It is enormous, but bounded. We'll leave to the more arcane academic journals the joys of deciphering the mathematics of *Finnegans Wake*, but I can't help mentioning, in the context of symbolic numbers, the novel's famous hundred-letter words, like this one: bababadalgharaghtakamminarronnkonnbronntonnerronntuonnthunntrovarrhounawnskawntoohoohoordenenthurnuk, which I'm sure you can tell is the sound of a thunderclap. Specifically, the one that reverberated around the heavens at the moment of the fall of Adam and Eve. There are ten of these "thunder words," but in fact they don't all have exactly one hundred letters. The first nine do, and then the final one has 101 letters, making a grand total of 1,001, another symbolic number with many cultural resonances.

Coming back to "Ithaca," Stephen departs as he entered, with geometry:

How did they take leave, one of the other, in separation?

Standing perpendicular at the same door and on different sides of its base, the lines of their valedictory arms, meeting at any point and forming any angle less than the sum of two right angles.

This is a conscious mangling of Euclid's fifth postulate: If a straight line falling on two straight lines makes the interior angles on the same side less than two right angles, the two straight lines, if produced indefinitely, meet on that side on which are the angles less than the two right angles.[6] If the men had remained parallel, as they were at the start of the chapter, then these interior angles would add up to 180°, which is two right angles, and the lines would not be able to meet. Or at least they wouldn't in standard Euclidean geometry. Joyce knew that kinds of geometry had been discovered in which the parallel postulate does not hold, but the setup has given us parallel lines, so the mathematical catechism has resulted in a contradiction—another in-joke from Joyce for the mathematically inclined.

For the writers in this chapter, mathematics is more than a way to communicate: it is a vital way of understanding the world. Mathematics has meaning, whether you are a village carpenter like Adam Bede or a deckhand like Ishmael. It is a refuge, a solace. Still, there are risks. Melville shows us the tragic outcome of assuming, like Ahab, that statistics give us complete control, and Joyce's absurdist calculations remind us that just because a number sounds impressive, that doesn't make it correct. The novels in this chapter have shown life through the prism of mathematics, from the smallest scale to the largest—from late-night rambles in Dublin to the entire sweep of human history. For these novelists, mathematics is the key.

7

Travels in Fabulous Realms

The Math of Myth

In Jonathan Swift's 1726 novel, *Gulliver's Travels*, the intrepid traveler Lemuel Gulliver visits the miniature land of Lilliput. He gives lots of detail about the precise dimensions of the people there, and he describes how the king of Lilliput arranges for Gulliver to be fed:

> *His Majesty's mathematicians, having taken the height of my body by the help of a quadrant, and finding it to exceed theirs in the proportion of twelve to one, they concluded from the similarity of their bodies, that mine must contain at least 1724 of theirs, and consequently would require as much food as was necessary to support that number of Lilliputians.*

While we do not judge satirical novels by the plausibility of their science, this is still an irresistible challenge. Where does this 1,724 come from, and is it correct? Spoiler alert: no, it's not, and if Mr. Gulliver is going to bring my Lilliputian colleagues' academic integrity into question with a howler like this, it's my duty as a mathematician to defend them. Previously in this part of the book, I've shown you how mathematics makes itself visible in fiction in several ways, from symbolic pattern numbers to lovely mathematical metaphors. In this chapter, we're going to explore another way that mathematics can be deployed: the narrative technique that I call *performative arithmetic*. As in the calculation above, it's often used when the narrator is recounting something that may appear unbelievable. A dash of solid fact, in the form of a mathematical calculation, gives plausibility to proceedings.

This is exactly what's going on when Gulliver visits the floating island of Laputa later in his travels. He gives us another calculation. The island is, he says, "exactly circular, its diameter 7,837 yards, or about four miles and a half, and consequently contains ten thousand acres." We readers can check this calculation for ourselves. An acre is 4,840 square yards, so ten thousand acres is a good approximation for the amount of land such a circle can contain—to the nearest whole number, it is 9,967 acres. The sleight of hand here is the elision between the verifiability of the arithmetic and the verifiability of the narrative. The mathematics is (roughly) correct, but that does absolutely nothing to establish that such a circular island exists. The spurious precision of 7,837 yards is probably designed to increase the illusion that this is a report of reality. It actually makes the calculation less accurate because the much rounder number of 7,850 would have given an area of 10,000 acres almost exactly—less than half an acre short.

In this chapter, I'll give you the tools to turn the tables on some literary logic and ask: Does this really stack up? We can check the working of the Lilliputian mathematicians and can laugh with Voltaire at the self-aggrandizing antics of the humans who turn out to be the tiniest of creatures in comparison to the giant visitor Micromégas, from his planet near Sirius. Are these fantastical lands possible, and what would life be like for their inhabitants? I'll show you the math that proves just how magical these creatures must be.

As Peter Pan tells Wendy, "You see children know such a lot now, they soon don't believe in fairies, and every time a child says, 'I don't believe in fairies,' there is a fairy somewhere that falls down dead." Not wanting to have any fairy slaughter on my conscience, I must emphasize that if anything I say makes it seem as if flying horses or giants or tiny people can't exist, all I mean is that if you do encounter one, something beyond our normal laws is occurring. As we'll see, creatures like the giant spiders that live in the Forbidden Forest at Hogwarts must be highly magical beings to defy all the mathematics that might otherwise "prove" they can't exist. Which is absolutely fine by me (as long as I don't have to be in a room with one).

I want to talk about giants first because my feeling is that, over the course of history, they have been taken more seriously than other fantastical beings as creatures that could potentially exist. There are several giants in the Bible, for example. In children's literature we meet Roald Dahl's beloved BFG, the half-giant Hagrid from the Harry Potter series, and many others. Giants have been popular in satirical novels, too. The French author François Rabelais (who gives us the adjective Rabelaisian, meaning "bawdy and crude") is most famous for his *Life of Gargantua and Pantagruel*, a five-volume work concerning two giants and their exploits. To give you just a hint of the cheerful tone of voice in which it's written, the full English title of the volume in which we first meet Pantagruel is *The Horrible and Terrifying Deeds and Words of the Very Renowned Pantagruel King of the Dipsodes, Son of the Great Giant Gargantua*. The exaggerated size of a giant emphasizes our own inescapable physicality, so it's a way to poke fun at our occasional coyness about ourselves. Rabelais enjoys being ridiculous. Gargantua (from whom we get the word "gargantuan") is born by climbing out of his mother Gargamella's ear, and it only gets sillier from there. The books are full of numbers and calculations about things like how much fabric is needed for Gargantua's codpiece (sixteen and a quarter ells, or about twenty yards, since you ask), but they are thrown about with gleeful abandon, much as we might now joke that something costs a million bajillion dollars.

None of the numbers given to describe the size of Gargantua make any attempt at consistency—it's all just exuberant silliness. We hear that the baby Gargantua's milk was supplied from a herd of "seventeen thousand nine hundred and thirteen cows of the towns of Pautille and Brehemond." For his shoes "were taken up four hundred and six ells of blue crimson-velvet, and were very neatly cut by parallel lines, joined in uniform cylinders." He combs his hair with a nine-hundred-foot comb whose teeth are entire elephant tusks. When Gargantua visits Paris and relieves himself in the street, he accidentally drowns "two hundred and sixty thousand four hundred and eighteen, besides the women and little children." In more numerical bawdiness, when Gargantua's wife dies, he thinks fondly of a certain "little" part of her anatomy "yet it had in circumference full six acres, three rods, five poles, four yards, two foot, one inch and a half of good woodland measure." This is all good fun, but because Rabelais doesn't tell us how big the giants are, it's pointless even to ask whether they could exist in real life, because we don't have enough information to make a reasonable assessment.

Let's visit Brobdingnag, then, because there we have very precise information. Brobdingnag, the land that Lemuel Gulliver visits after Lilliput, is a kind of inverse to that country, because everything in Brobdingnag is twelve times as big in every dimension as in our world. This is rather convenient in that it means anything that would normally be an inch long (a wasp, for example) is now a foot long. So it's not just the people that are giants, but the plants and animals too, and even the weather. On one occasion, Gulliver is unlucky enough to be caught outside in a hailstorm: "I was immediately by the force of it, struck to the ground: and when I was down, the hailstones gave me such cruel bangs all over the body, as if I had been pelted with tennis-balls. . . . Neither is this at all to be wondered at, because nature, in that country, observing the same proportion through all her operations, a hailstone is near eighteen hundred times as large as one in Europe."

Where does this "near eighteen hundred" come from? Well, we know that every dimension is multiplied by 12. So the hailstones are twelve times as long, twelve times as wide, and twelve times as high as ours. This means their volume is not twelve times as much, but $12 \times 12 \times 12 = 1{,}728$ times as much, or, "near eighteen hundred" (though really it's nearer seventeen hundred). This is the start of the problem with giants. If you scale something up in every dimension by the same factor—here it's 12, but let's say it is some fixed k—then the volume would change by a factor of $k \times k \times k$, which we write in mathematical notation as k^3, because there are three k's multiplied together. In other words, the volume changes with the *cube* of the scaling factor. Meanwhile, any area connected with the object will only change with the *square* of the scaling factor. To see what I mean, have a look at the diagram below. I've shown what happens to a box if we enlarge it in each dimension by a factor of 2. Imaginatively enough, it has width w, depth d, and height h.

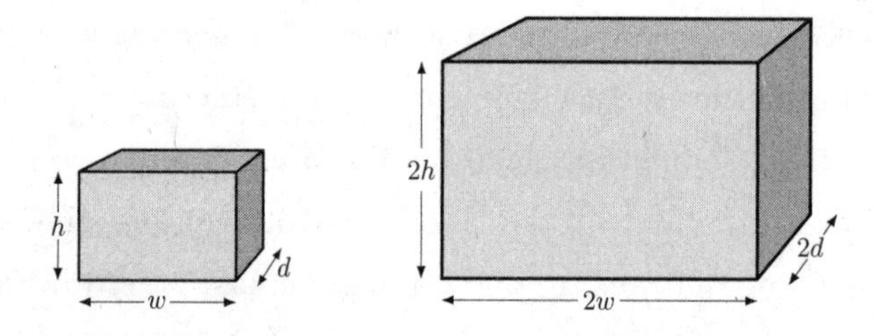

That means the box has volume V, where $V = w \times d \times h$. Now, if we enlarge by a factor of 2, the new, bigger box has width $2w$, depth $2d$, and height $2h$. That gives it a volume of $2w \times 2d \times 2h = 8(w \times d \times h) = 8V$. So yes, this agrees with our reasoning, because $8 = 2^3$. On the other hand, the area A of the base of the original box is $w \times d$, but the doubled box has base area $2w \times 2d = 4A$, and $2^2 = 4$.

I said that this squaring factor was true of *any* area connected with the object. What I meant by that is that it's not just the base area that increases by the square of the scaling factor, but, for example, the area of any cross section through the box, and also its surface area, that has this property. We don't need to work out the exact formula for the surface area to know this (but if you want to, it's $2(wd + dh + wh)$); it's enough just to realize that the calculation involves adding together a bunch of areas that involve two of the measurements multiplied together, and so doubling each of the measurements will multiply the total by 4. In the more general case, the box enlarged by a factor of k will have volume k^3V and area k^2A. This fact is known as the *square-cube law*.

Here's where things get nasty for giants. When humans are moving around, the weight of their bodies must be supported by their skeletal structure. Studies show that the human femur (thigh bone) will break under about ten times the pressure it normally has to carry. You might remember from high school science that pressure is the force per unit area. That is, $\text{pressure} = \frac{\text{force}}{\text{area}}$. The area here is the cross-sectional area of the femur. The force exerted comes from our mass being pulled down by gravity, and our mass is roughly proportional to our volume. All this means that the pressure on our femurs is proportional to $\frac{\text{volume}}{\text{area}}$. Now, if we scale up the human body by a factor of k, the square-cube law tells us that the volume increases by a factor of k^3, but the area increases by a factor of only k^2. Putting this all together shows that the pressure on the scaled-up human's bones will be proportional not to the original $\frac{\text{volume}}{\text{area}}$, but to $\frac{k^3 \times (\text{volume})}{k^2 \times (\text{area})}$, which, if we cancel a couple of k's, is $k \times \frac{\text{volume}}{\text{area}}$. In other words, the pressure on the scaled-up human's bones is k times the pressure on our bones. The Brobdingnagians are twelve times the size of Gulliver. That means the pressure on their bones, just standing still, is twelve times the pressure on his bones. But bone can take only ten times normal pressure before breaking; the Brobdingnagians' bones would break as soon as they tried to move.

So the Brobdingnagians cannot really exist. The same is unfortunately true for the BFG, and for Giant Pope and Giant Pagan of John Bunyan's *Pilgrim's Progress*, who are an estimated sixty feet tall, ten times the height of the hero, Christian (unless divine intervention is involved—all bets are off with an omnipotent deity). King Kong, if he could exist at all (the films are far from consistent about the size he is supposed to be), would be incredibly weak—he would barely be able to support his own weight, never mind jumping around on skyscrapers and punching airplanes out of the sky. Fay Wray could probably beat him in a fight!

There is some hope, though, for slightly smaller giants. In the Harry

Potter books, Rubeus Hagrid, Keeper of the Keys at Hogwarts School of Witchcraft and Wizardry, is a half-giant. He's described as being twice the normal height but, crucially, three times the normal width. Assuming that he's also three times the normal depth, that would mean the cross-sectional area of his bones is 9 (or 3 squared) times ours, but his mass is only 18 times ours, not 27 times. That would mean the pressure on his bones is twice what it is on ours. He could definitely still walk around, and maybe even run, but he would likely be prone to broken bones, and he should certainly not take up skipping. The same goes for King Og of Bashan, a biblical giant whom Moses meets in the book of Deuteronomy. His exact dimensions aren't given, but we are told that his bed was 13 feet long, so he was perhaps scaled up by a factor of two, thereby doubling the pressure on his bones. Again, survivable, but he wouldn't be a mighty warrior.[1]

Before we move on, I want to tell you about *Micromégas*. It's a short satirical novel by Voltaire, which I am both grateful and annoyed to have heard about accidentally. I'm one of those people whose brain won't shut up even when I want it to, and so I often listen to an audiobook or the radio to drift off to sleep. Of course, it can't be anything too exciting. The audiobook company Audible, realizing that lots of people do this, released a series of "Bedtime Stories for Adults" that were deliberately not very exciting. I am appalled to have to tell you that the second of these was *A Short Account of the History of Mathematics*, by W. W. Rouse Ball. How dare they! In with this, and a book on quilt collecting, was *Micromégas*, by Voltaire, which I'd never heard of. Imagine my delight when it turned out to be the story of a giant named Micromégas, from a planet orbiting Sirius, who visits Earth. What's more, Voltaire describes his exact size and talks about the calculations that mathematicians can make to determine the size of his home planet. So, Monsieur Voltaire, you brought mathematicians into this. Let's see how your calculations fare.

Micromégas is a satire on human vanity and pomposity. We puny beings chatter among ourselves, thinking ourselves important, but we are as ants compared to Micromégas. He is so huge he can't even see us. He visits our solar system, first meeting the people of Saturn, who are 6,000 feet tall, before traveling to Earth. He can communicate with humans, whose voices are ridiculously tiny, only by means of an ear trumpet that, for reasons best known to himself, he makes from his own fingernail clippings. Now, Micromégas is described, essentially, as a human scaled up by a factor of 24,000. The square-cube law tells us that the pressure on his bones is therefore 24,000 times the pressure on ours. He would immediately collapse under his own weight on Earth, as indeed would his Saturnine friends. But this got me wondering whether perhaps on a different planet, with different gravitational forces, giant humanlike beings could in fact exist.

Voltaire writes:

> *Certain geometers,*[2] *always of use to the public, will immediately take up their pens, and will find that since Mr. Micromégas, inhabitant of the country of Sirius, is 24,000 paces tall, which is equivalent to 120,000 feet, and since we citizens of the earth are hardly five feet tall, and our sphere 9,000 leagues around; they will find, I say, that it is absolutely necessary that the sphere that produced him was 21,600,000 times greater in circumference than our little Earth. Nothing in nature is simpler or more orderly.*

So let's have a look at this calculation, shall we? I suspect that Voltaire is teasing these geometers a little for being so certain about everything. What is claimed is that since Micromégas is 24,000 times our height, then his planet must be 24,000 times the circumference of ours. So it should be 9,000 leagues multiplied by 24,000. Already there's a problem, because 9,000 times 24,000 is 216,000,000. Sorry, Voltaire, but you are a factor of 10 off. A bigger issue is whether it is right even to make this deduction. Would giants be expected to come from a giant planet? The answer has to involve gravity. Remember that when we talked about the pressure on our bones, I said that the force comes from gravity acting on our mass. If you visit a planet where gravitational pull is twice what it is on Earth, then the pressure on your bones will double. We can work this in the other direction and say that if Micromégas comes from a planet where the pressure on his bones matches the pressure on humans' bones on Earth, that planet must have gravity 1/24,000th of Earth's gravity.

Can such a planet exist? What's the gravity on an Earth scaled up by a factor of 24,000? The laws of gravity have been known since Isaac Newton: the force of gravity obeys what's called an *inverse square law*. This means that the pull of gravity exerted on you by an object (say, the earth) depends on 1 over the square of your distance from that object. If your distance from the center of the earth doubles, then the force of gravity would be divided by the square of 2—that is, 4. But wait, you say, surely that would mean that gravity would be lower at the top of Mount Everest than it is at sea level? And yes, that's absolutely true. At the top of Mount Everest, gravity is 9.77m/s^2, while on the surface of the Arctic Ocean, gravity has been measured at 9.83m/s^2. So, if you want to lose weight, forget the diet; just move to higher ground.

The other thing that affects gravity on a planet is the mass of that planet. If the mass doubles, then gravity doubles. We can encapsulate all this in a simple expression:

$$\text{planet gravity} \propto \frac{\text{planet mass}}{(\text{planet radius})^2}$$

That symbol $\propto$ means "proportional to." It's saying that while these things aren't exactly equal, they vary exactly in line with each other. If the right-hand side doubles, so does the left. So, imagine we do the simplest possible thing and just double Earth. Its volume, and hence its mass, will increase with the cube of the scaling factor, so the mass will increase by a factor of 8. But at the same time the distance from the center to the surface—in other words, the radius—will also double, so the square of the radius increases with the square of the scaling factor. Thus, the square of the radius increases by a factor of 4. This is just like the square-cube law! The net effect is that gravity on the double-Earth planet is going to double.

If we had an Earth-like planet whose radius was 24,000 times ours, its gravity would also be 24,000 times ours. So this would be even worse for Micromégas than Earth. To get the same pressure on his bones as we experience on our Earth, his planet, if it's like ours in everything but size, would have to be not 24,000 times the size, but 1/24,000th the size. Micromégas could not live on a planet 1/24,000th the size of Earth. He is 120,000 feet tall, and the circumference of such a planet would be only 5,478 feet. It would be like a human trying to live on the surface of a grape. Voltaire's geometers were totally wrong. What we will never know is whether Voltaire was aware of these two errors. He may have introduced one or even both of them on purpose just to show that even such clever people as geometers are fallible—this is a satire on pomposity, after all. Alternatively, his arithmetic might just not have been very good.

We could go further and wonder about more sophisticated scenarios—planets that are not earthlike, for instance. We could get the same gravity on a bigger planet if it was less dense, because while the mass of a planet does depend on its volume, it also depends on its density. Perhaps the huge Sirian planet has a lower density. The least dense planets that we know of are known, rather whimsically, as *super-puff* planets. They have a density about 1 percent of Earth's, which means the best we can hope for on the Sirian planet is 240 times our gravity. It's still hopeless. Life-forms on other planets are extremely unlikely to resemble humans, of course. Next time you are reading a science fiction novel in which aliens invade Earth, you'll be able to cast your expert eye over the descriptions of these aliens and get a fairly good idea what size planet they must come from, and whether their terrestrial sojourn is going to result in any broken alien legs. The mathematics of the square-cube law can give us some useful ground rules.

. . .

The square-cube law is bad news for the Goliaths of this world, but it also raises some questions about animals. The smallest mammal is the bumblebee bat, which weighs a tiny 1.7 grams (less than a sixteenth of an ounce) and is about an inch and a quarter long. Compare this to the gargantuan (thank you, Rabelais, for the adjective) blue whale, which has been recorded at sizes of up to 98 feet long and weights over 200 tons. How can this be? Big animals are not, and cannot be, small animals simply scaled up, because they would be crushed under their own weight. The answer is evolution. Think about the legs of a mouse compared to those of an elephant. The elephant's legs must be proportionately much thicker because the cross-sectional area of its bones has to keep pace with the increase in volume, and hence mass. The square-cube law was first observed by Galileo (though he did not give it a name) in this context.

We know that mammals can evolve to be very big, but fiction has been fascinated with other giant creatures, such as the six-foot cockroaches infesting the New York subway in the 1997 Guillermo del Toro movie *Mimic*, based on a short story by the American science fiction author Donald A. Wollheim. And who can forget *Them!*, a 1954 movie featuring giant ants rampaging through the New Mexico desert? Or, as the original poster for the movie had it: "A horror horde of crawl-and-crush giants clawing out of the earth from mile-deep catacombs!" Their unnatural size turns out to have been caused by radiation from testing the atomic bomb. Are giant insects possible? How about the elephant-sized acromantula Aragog, a magical giant spider living in the Forbidden Forest of Hogwarts?

The heaviest adult insect we know of is the aptly named giant wētā, which can grow to a hideous 8 inches long and a weight of over 2.5 ounces. Meanwhile, the grubs of the Goliath beetle can be even heavier, at an appalling 4 ounces (though shorter at 4.5 inches). Stick insects can be longer but are much lighter because of their sticklike shape. The longest insect ever recorded was a giant stick insect at the Insect Museum of West China. It was an unforgivable 25 inches long. Spiders can grow even bigger. The Goliath birdeater is the heaviest spider known to science, at 6.2 ounces, with a length of 5.2 inches. I apologize that you now know that. At this point we may be quite keen for mathematics to step in and prove that insects and arachnids can't get even larger than this. Forgive my anti-bug prejudice, by the way. I like a butterfly or bumblebee as much as the next person, but I draw the line at giant beetles, however essential they are to the ecosystem. Happily for any other bugophobes around, there are some natural limits, in spite of what evolution can manage in terms of adaptations like thicker legs and so on.

The first issue is that insects and arachnids have their skeleton on the outside of their body (exoskeleton). While this gives them a strong structure, it means that they have to shed their skin, usually several times, as they grow. While the new skin is hardening, it leaves them not only vulnerable but very weak. Above a certain size, periodically losing your exoskeleton just doesn't work.

Another factor at play is oxygen. Like all animals, insects and arachnids need oxygen to survive, and it must reach every part of the body. Larger animals, like mammals and birds, have circulatory systems in which blood, pumped by the heart, carries oxygen around the body in blood vessels. Oxygen gets into our bodies through the lungs. It is absorbed through the surface of the lung, which for this reason has many tiny folds to maximize surface area. According to the American Lung Association (and they should know), the total surface area of your lungs is about the size of a tennis court, and if you laid all the airways running through your lungs in a long line, it would stretch fifteen hundred miles.

Now, insects and arachnids don't have lungs. They have a substance like blood, but it does not transport oxygen. Instead, they absorb oxygen directly from their body surface, and it travels to their cells through tiny tunnels called *tracheae*. And here's where the square-cube law comes in. The amount of oxygen an insect can absorb is proportional to its surface area. But the amount it needs is proportional to the number of cells it has, which grows with its volume. As we know, surface area depends on the square of the scaling factor, but this will soon be outpaced by the fact that volume depends on the cube of the scaling factor. Our giant insects would suffocate.

Just to reassure ourselves, let's work out an approximate maximum size for a spider. Research at the University of California, Irvine, in 2005 showed that insects can actually cope with concentrations of oxygen just one-fifth of what is in our atmosphere. According to the square-cube law, if we scale up our bug by a factor of k, then each square centimeter of surface area needs to supply k times more oxygen. Therefore, in theory, the upper limit of growth is five times, before the bug suffocates. Our friend Aragog is said to be elephant-sized, about two to three meters long. Even a ten-times-scaled Goliath birdeater would be only 4'4", so Aragog must definitely be magical. If we use a five-times-scaled Goliath birdeater as our maximum, then we are looking at a spider around two feet long. I still wouldn't want to bump into one of those.

Any paleoentomologists reading this will be protesting at this point that prehistoric insects grew much bigger (shudder). For example, there were species of very large dragonfly-like creatures called Meganisoptera. Among the largest of these were *Meganeura monyi*, which were flying around Europe in the late Carboniferous period, about 300 million years ago, before even the dinosaurs came along. One fossil specimen had a wingspan of 2'4" and an estimated mass of over seven ounces. How is this possible? One factor is that at various points in prehistory, oxygen concentrations in the atmosphere were much higher than they are now. This increases what can be absorbed through the exoskeleton. But perhaps more important is predation. When insects had the skies to themselves, they could grow very large with impunity. But once pterosaurs came on the scene, things changed. A two-foot dragonfly makes a delicious meal for a pterodactyl.

Just as insects and arachnids don't benefit from being large, warm-blooded creatures find it difficult to be tiny. That's probably why insects and arachnids have evolved to be smaller, thus filling a niche that mammals can't. Unlike insects, warm-blooded creatures lose heat from the surface of our bodies, so our heat loss is proportional to our surface area. As we shrink, the amount of heat our bodies produce decreases much faster than our surface area. Small mammals lose heat much more quickly compared to the amount of heat they generate, and at some point they just can't maintain their body temperature. There are various evolutionary tricks to deal with this. Small mammals usually have a much more spherical shape than larger ones, and are covered in fur (compare a mouse with a rat—the mouse is fluffier and rounder in shape). We also don't find small mammals in colder climates. We get Arctic hares, but not Arctic rabbits. Mammals are of course smallest when they are born, and they often have fur or other adaptations to help them stay warm during this phase—think of "puppy fat." It's the same with birds—those adorable little newborn ducklings are much fluffier than their parents. At the other end of the scale, overheating can be a real problem for the largest mammals, whose huge bodies produce more heat than their proportionally smaller surface area can easily dissipate. Adaptations for this issue include things like the large ears of an African elephant.

It seems that colossi cannot bestride the earth (at least not without supernatural help of some kind). But what about that other staple of fairy tales and fables: tiny people? We've already encountered the Lilliputians, and what's more, we know their exact size, thanks to Lemuel Gulliver. Shrink rays, miniaturizing potions, or mysterious radioactive fogs have all caused movie characters to become various levels of tiny. The eponymous *Doctor Cyclops* shrank his terrified victims to a height of twelve inches in the 1940 movie, while *The Incredible Shrinking Man* (1957) is doomed to continue shrinking forever. More recently, the hapless inventor in *Honey, I Shrunk the Kids* zaps his children to a quarter of an inch in height—a tremendous shrinking factor of about two hundred, while Matt Damon shrinks to five inches tall in the 2017 movie *Downsizing*.

Fairies, pixies, and other fantastical creatures are small, but they aren't specifically miniature humans, nor, usually, are their exact sizes given, so it's hard to deduce much about their physical properties. One of my daughters once required me to make a dress for the tooth fairy, to be left out with the tooth. From that I can tell you that (a) I'm not a very good seamstress, and (b) the tooth fairy is two inches tall; but (c) if she fits into that dress, she definitely does not have standard human proportions.

We do have slightly more information about literature's best-loved tiny family. Pod and Homily Clock, and their daughter Arrietty, are Borrowers in the popular series of children's books by Mary Norton. The Borrowers are essentially small versions of humans, an estimated one-sixteenth of our size, who live in human houses in out-of-the-way places. The Clocks live under the grandfather clock in the hall of Aunt Sophy's house. They live by "borrowing" things they need—needles, safety pins, matchboxes, buttons, bits of paper, spools of thread—all those little things you can never find when you need them, and now you know why. You might have seen the charming Studio Ghibli movie *Arrietty*, which adapts the first Borrowers novel.

What would life be like for these little people? I'm going to focus mainly on life in Lilliput, because Gulliver tells us the exact dimensions of the Lilliputians and their world (a factor of twelve smaller than ours), but you can play around with the same ideas in other worlds. When Gulliver is shipwrecked on his travels, he washes up on the shores of Lilliput and, after initially being pinioned to the ground by the Lilliputians, is accepted into their society, giant though he is. He even assists them in the war between Lilliput and neighboring Blefuscu, these two realms representing, respectively, Britain and France. The issue over which the war is fought is as small as the creatures fighting it: in Lilliput, tradition dictates that you open your soft-boiled eggs at the little end, whereas the Blefuscans are Big-Endians, and naturally this insult to decency cannot stand. We are invited to witness the ridiculousness of these arguments and compare them to our own petty concerns.

The first thing to say about life in a small country is that Lilliputians benefit hugely from the square-cube law when it comes to strength. Remember that when we talked about how much pressure is on the bones of a scaled human, we found that if you scale by a factor of k, the pressure on the bones also scales by a factor of k. Here, we are scaling by a factor of 1/12, and so the same thing happens to the pressure on Lilliputians' bones. They would be comparatively much stronger, able to carry many times their own weight. In stories, we sometimes see tiny people in peril because they are in a high place, like on top of a human table or being carried on Gulliver's shoulder. The fall from such a height, while negligible to us, would surely be fatal to a Lilliputian, right? Well, there's a curious thing about falling. The reason it's dangerous is that as we fall, kinetic energy builds up, which is all released very suddenly when we hit the ground. But we don't keep on accelerating indefinitely. You may have heard the phrase "terminal velocity" in this context. As we fall, we accelerate under gravity, but there is a small opposing force upward due to air resistance. This air resistance is proportional to the speed we are moving at but also to the area in contact with the air. As our speed increases, the air resistance increases, until at some point the two forces (gravity and air resistance) balance out, and at that point we stop accelerating—we are at terminal velocity.

Human terminal velocity is a pretty conclusive 50 meters per second. It's been calculated by NASA that we have a good chance of walking away from an impact at up to 12 meters per second, but anything higher than that risks serious injury or death. The reason parachutes work is that they increase our area, and this increases the air resistance, meaning the equilibrium point is reached sooner and the terminal velocity is therefore lower. So what's the terminal velocity for a scaled-down human? Well, the downward force due to gravity is proportional to our mass, which is proportional to our volume. And air resistance is proportional to our surface area. This means that if we are scaled by a factor of k, the downward force due to gravity is scaled by k^3 and the upward force due to air resistance is scaled by k^2—the square-cube law strikes again! This means that these forces will now match up only at k times the original terminal velocity. For our Lilliputians, the scaling factor is $k = \frac{1}{12}$, so their terminal velocity is just one-twelfth of ours, a mere 4.2 meters per second.

Okay, so we are happily falling through the air. What happens when we hit the ground? All that kinetic energy we have built up has to be dissipated. I did a few calculations and it turns out that the maximum survivable velocity of a human scaled by a factor of k is proportional to $\frac{1}{\sqrt{k}}$. Now, humans can survive 12 meters per second. So the scaled humans can survive $12 \times \frac{1}{\sqrt{k}}$ meters per second. If you plug $k = \frac{1}{12}$ into this, you find that Lilliputians can survive impact at a velocity of $12\sqrt{12}$, or about 42 meters per second. But hang on—their terminal velocity is just 4.2 meters per second. That means that whatever height they fall from, their speed will never exceed 4.2 meters per second, and so they can easily survive a fall from any height. No need to rappel down Gulliver's legs—they can just jump off his head and be completely fine. The scientist J.B.S. Haldane made a similar point about animals falling in a 1927 essay, "On Being the Right Size," with an arresting metaphor. He said that you can drop a mouse down a thousand-yard mine shaft and it will be fine. But a man is killed, and a horse, says Haldane, "splashes."[3]

Another terrifying situation for shrink-ray victims is getting trapped in some vast receptacle, like a jam jar. But this again is no problem. In fact, to jump to a particular height uses an amount of energy roughly proportional to your mass. Meanwhile, the amount of energy that muscles produce is also roughly proportional to their mass. That means the scaling factors cancel out here, and the height a scaled-down human can jump to is more or less the same as the height a usual human can manage, around a meter, unless you are an adept high jumper. So our Borrower in a jam jar can simply hop out without difficulty. Incidentally, this also shows the silliness of those claims you sometimes see that if a flea were the size of a person, it could jump over a skyscraper. In fact, before our unfortunate giant flea suffocated and collapsed under its own body weight, it could jump to about the same height as its standard-sized brethren, about seven inches.

So far, then, everything is looking pretty good for our Lilliputians. But there are some downsides. I have already pointed out that small mammals have a lot of work to do to keep warm. Because they lose heat proportionately much faster than us, getting cold will be a serious risk. A few years after Gulliver visited Lilliput, this heat loss phenomenon actually changed the life of a young instrument maker in eighteenth-century Glasgow. He was asked to take a look at the university's scale model of the famous Newcomen steam engine, a very early steam engine designed by Thomas Newcomen, widely used for pumping water out of mines. The engine worked by repeatedly heating and cooling a cylinder—the cooling condensed the steam, creating a partial vacuum that made the piston move. The engine did work, but it wasn't very efficient because a lot of heat energy was lost by this repeated temperature change. The scale model, though, didn't work at all, and the mysterious thing was that it was very accurately done, just like the full-size engine, except smaller. Now that you are an expert on the square-cube law, you can perhaps spot the problem. The heat loss that made the real engine somewhat inefficient was hugely magnified in the model because heat loss depends on surface area and heat production is proportional to volume—just as in animals.

In trying to create a working model, the ingenious instrument maker came up with the idea of a separate condenser, which was a groundbreaking development in steam engine design and a major contributor to the Industrial Revolution. James Watt—for that was the young man's name—became famous, and the scientific unit of power, the watt, is named for him. All because of the square-cube law.

I don't know whether the Lilliputians had steam engines, but frankly that is the least of their worries. I'm afraid there is bad news concerning their metabolic rates. Gulliver reports the calculation of the Lilliputian mathematicians that because he is twelve times as big as they are in all dimensions, he will need 1,724 times as much food as they do. The reasoning behind this is presumably that because his mass is 12^3 times theirs, his energy needs are 12^3 times theirs as well. This number is actually 1,728 (readers of a certain vintage may have dim memories of learning this number in school because it's the number of cubic inches in a cubic foot). Some editions of *Gulliver's Travels* do in fact correct the 1,724 stated in the text to 1,728. We'll never know whose error it was originally—the Lilliputian mathematicians? (Fie! Surely not!) Gulliver's faulty recollection? Jonathan Swift's arithmetic? It could just be a simple printer's error. If I had to choose one of these, unfortunately I'd have to plump for Swift's arithmetic.

I already mentioned his description of Laputa: exactly circular, with a diameter of 7,837 yards and an area of ten thousand acres. As I said, this 7,837 may be precise, but it's not accurate. To be fair to Swift, it's not a particularly easy calculation to do by hand—you have to convert acres into square yards, then divide by π, take the square root of that to get the radius of the circle, and then double it to find the diameter. Thank goodness for the calculator app on my smartphone. I shall magnanimously forgive Mr. Swift, even as he caricatures mathematicians by saying that the citizens of Laputa are so obsessed with mathematical speculations that they don't look where they are walking and to prevent injury need to employ servants to hit them with a bag of pebbles carried for the purpose, to distract them from their reveries. How silly to imagine that mathematicians can get distracted . . . anyway, what was I saying? Ah, yes, even the 1,728 number is not quite right, because we now know that there is a slightly more complicated relationship between the size of an animal and the amount of energy it uses.

Warm-blooded animals, like humans and other mammals, lose heat from their bodies at a rate proportional to their surface area. But animals use energy on other things too—keeping organs going, pumping blood, digesting food, and so on, and we might expect the amount of energy required for that to be approximately related to the mass of the animal. This means the probable amount of energy needed—the metabolic rate of the animal—is going to depend on both its surface area and its mass. The mass m of an animal of a given shape is proportional to the cube of its height, and the surface area is proportional to the square of its height. So if the metabolic rate were entirely due to heat loss (surface area), it would depend on the square of the cube root of height, or $m^{2/3}$, whereas if it were entirely due to keeping organs going, it would depend directly on the mass.

A Swiss scientist named Max Kleiber investigated mammals of different sizes in the 1930s, and he found that, to an impressive degree, the metabolic rate of a mammal is proportional to its mass m raised to the power ¾. What this means is that if we know a particular mammal requires 100 calories a day to survive, then an animal twice the mass will require not $2 \times 100 = 200$ calories but $2^{3/4} \times 100$, which is about 168 calories. This rule of thumb is now known as *Kleiber's law*. Current dietary guidelines say that an adult male, like Gulliver, needs about 2,500 calories a day. The mass of a Lilliputian mini-Gulliver is just $\frac{1}{1{,}728}$th that of Gulliver. So Kleiber's law tells us that mini-Gulliver will need $\left(\frac{1}{1{,}728}\right)^{\frac{3}{4}} \times 2{,}500$ calories a day. This works out to a puny 9.3 calories. So far, so good.

But there's a huge problem. As I mentioned before, in Lilliput everything is at 1/12 scale, not just the people. The trees, the crops, the livestock, everything is dollhouse size. But as anyone who has ever been on a diet knows, the number of calories in a given food is calculated from its mass. A hundred grams of sugar has twice as many calories as fifty grams of sugar, sadly. This means that the mathematics of Lilliputian agriculture is just not going to add up. To see what I mean, let's use the handy fact that an apple has approximately one hundred calories. Gulliver could get his calorie intake, then, from twenty-five apples each day. Now let's work out how many mini-apples our mini-Gulliver would need. His little apples are, again, $\frac{1}{1{,}728}$th the mass of normal apples. This means that each one contains just $\frac{100}{1{,}728}$ calories, which is a tiny 0.058 calories. The consequences of this are serious. To get his required daily calories, mini-Gulliver would have to eat 161 Lilliputian apples.[4] This is more than six times as many normal apples as Gulliver would have to eat. Mini-Gulliver would have to spend all day picking and eating apples! Imagine having to eat twenty-five meals a day—I'm no economist, but I think the farming industry would struggle to cope, and there might not be much time for the finer things like culture and waging war over which end to open your eggs.

The final challenge in Lilliput is water. You see, all liquids have surface tension; it's what allows things like raindrops and bubbles to form. Different substances have different amounts of surface tension, but it is an intrinsic property of a given liquid that, like density, is not affected by scale. If you immerse any object in water, it will come out covered in a thin film of water, about half a millimeter thick. This is why we have towels. Crucially, this half millimeter depends only on the surface tension and adhesive properties of water and not on the size of the object. If the body surface area of an average adult is about 1.8 square meters, the weight of the water you carry out of the bath is about 2 pounds. The average adult weighs 165 pounds, so 2 pounds added to this is not an issue.

The problem is that for a Lilliputian, their surface area, being an area, changes with the square of the scaling factor, so because 12 squared is 144, the weight of water they carry will be $\frac{1}{144}$th of what we carry, about a quarter of an ounce. Unfortunately, an adult Lilliputian's weight depends on their volume and so varies with the cube of the scaling factor. Consequently, a Lilliputian will weigh around 1.5 ounces. Suddenly the water is 14 percent of their weight, which would be like us putting on a 23-pound coat. Swimming would be a lot more tiring for Lilliputians than it is for normal-sized humans. I wouldn't want to be the children in *Honey, I Shrunk the Kids* either. They shrink by a factor of 200, which means that being immersed in water would be fatal to them—they would be surrounded by a wall of water twice their own body weight and would drown.

Meanwhile, a Lilliputian getting caught in the rain would be in for quite an ordeal. The size of raindrops is determined by the surface tension of water, so it would have to be the same in Lilliput. This means each raindrop would weigh about one-sixth of a percent of the Lilliputian's weight. This doesn't sound like much, but it would be like us being pelted with baseballs. It's a racing certainty that fairies, pixies, and elves (if they are small ones rather than the man-sized elves of Middle-earth) will do all they can to avoid the rain.

I can't quite bear to leave these considerations without mentioning the drinking habits of hobbits. In Middle-earth, the imagined world of J.R.R. Tolkien's *Lord of the Rings* trilogy, hobbits like Bilbo Baggins are around three feet six inches tall, essentially like humans except for their furry feet and slightly pointed ears. One scene in Peter Jackson's film of Book 1 of the trilogy has a hobbit excitedly discovering that beer in the human pub in the village of Bree comes in vast measures called "pints." Now, hobbits aren't that much smaller than humans, so we might think the effect of a pint wouldn't be that much different. But given that the effect of alcohol is roughly proportional to your volume, we have to cube the scaling factor. When you do this, you find that a pint of beer will have the same effect on a hobbit as five pints will have on a man. They had better stick to halves!

In the first part of this book, we uncovered the hidden mathematical structures of literature. In this second part, mathematics has become visible in the words and allusions of writing. We've seen that even the numbers we encounter in stories have a symbolism deeply rooted in mathematics. There are good mathematical reasons why there are three wishes, seven dwarfs, forty thieves, and a thousand and one Arabian nights. Mathematical ideas themselves are crafted into wonderful metaphors by writers like George Eliot and Herman Melville. Actual calculations can also be pressed into service. James Joyce uses them both to reveal and to obscure. Leopold Bloom's budget is telling in its omissions, while the dizzying permutations of Dedalus's and Bloom's relative ages appear to make sense but do not.

In this chapter, we've seen how authors like Jonathan Swift and Voltaire use calculations in a different way, humorously using our instinctive trust in the "truth" of mathematics to lend an air of authority to their fantastical stories. We've kept the receipts, though, and that lets us poke a little affectionate fun, in our turn, at the prospect of rampaging giant insect hordes or miniature civilizations.

Mathematical symbolism and metaphor are present in every kind of literature, from the humblest of fairy tales right through to *War and Peace*. They are right there waiting to be discovered—and now you have the tools to find them.

Part III

Mathematics Becomes the Story

8

Taking an Idea for a Walk

Mathematical Concepts So Compelling They Escape into Fiction

Every so often, a mathematical idea grabs the public imagination. In the twentieth century, mathematical hot topics such as fractals and cryptography were key plot features in novels, though not always in an entirely accurate manner. (If anyone feels like setting up a Bad Math Award, there are plenty of contenders.) In the nineteenth century, the mysterious new "fourth dimension" was all the rage. *Flatland* (1884), a bestselling book by Edwin Abbott, used the ideas of two, three, and four dimensions to satirize Victorian values, and it has spawned numerous spin-offs and sequels since then. The protagonist of *Flatland* is a living, breathing embodiment of geometry, in the form of a square, and much of the plot revolves around the mathematics of dimensionality.

This final part of this book shows mathematics coming into the spotlight. We have built the foundation of our house of literature with mathematical structures, we have furnished it with mathematical metaphors, and now we're ready to populate our house with mathematical characters, ideas, and people. In this chapter, I'll show you how the mathematics that made it out of the textbooks and into the popular consciousness has been treated in fiction, not just with occasional number-related metaphors (or "figures of speech," as these should clearly be called) but as an integral part of the narrative.

We'll take a tour around *Flatland* and meet its curious polygonal inhabitants, and then we'll see how other authors have plotted a course to higher dimensions.

Edwin Abbott Abbott was a teacher, clergyman, and author. For most of his career he was headmaster of the City of London School, a boys' school that he had himself attended as a child. Some years after Abbott's time, a sister school was set up, the City of London School for Girls, which I attended between 1988 and 1993. Another link between us is that one of the people who taught Abbott mathematics was Robert Pitt Edkins, who from 1848 to 1854 was Gresham Professor of Geometry and thus my academic ancestor. Nothing would please me more than learning that, like him, I may have inspired future mathematical fiction writers.

Edwin Abbott was well known in his time not only as a brilliant teacher and headmaster but also as a respected thinker and author. He wrote more than fifty books on theology and education, in particular the teaching of English and Latin, including such titles as *Handbook of English Grammar* (1873), *Oxford Sermons Preached Before the University* (1879), and that racy 1893 page-turner *Dux Latinus: A First Latin Construing Book*. In such company, 1884's *Flatland: A Romance of Many Dimensions* is rather a surprise.

The events of *Flatland* are narrated by "A. Square," a respectable member of Flatland society. His universe is entirely two-dimensional, a flat plane whose inhabitants are geometric figures.

In the first part of the book, A. Square describes Flatland, satirizing some of the worst aspects of Victorian society—the rigid class structure, the restrictive perception and treatment of women, and religious dogmatism in the ruling priestly class. In Flatland, the men are all polygons (triangles, squares, and so forth), while women are lines. They are one-dimensional beings in a two-dimensional world, incapable by their very nature of any kind of equality with men. (I must stress that this absolutely was not Abbott's view. He was a strong proponent of improving access to education for girls and women, and he was in contact with several prominent women who also supported this cause, including George Eliot.) Moving around Flatland, a woman viewed end-on is just a point—almost indiscernible. This can be very dangerous to the men—you can be impaled accidentally by a careless woman, and of course they are all careless. For this reason, whenever women leave their houses, they must maintain a constant "peace cry" to warn others of their presence. Some regions also insist that women constantly wiggle their back ends from side to side, or that they be accompanied by a man if leaving the house. Houses in Flatland are regular pentagons, because square houses have sharp right-angled corners that pose a health risk to anyone accidentally bumping into them. There are separate male and female entrances, also for safety reasons: we don't want any men to be accidentally eviscerated by their wives coming out of the house as they are going in. Here is a diagram from the book, showing a typical house:

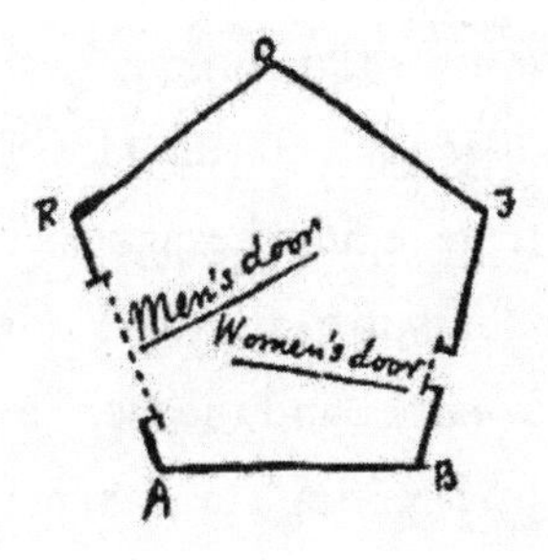

Note the dreadful pun—the sides RO and OF together make the ROOF.

For the male population of Flatland, regularity and symmetry are all-important markers of status. At the bottom of the pile are the poor benighted isosceles triangles. These degenerates can do damage with their sharp angles, and thus the more obedient among them can be used as soldiers. However, there is some chance of upward mobility:

"After a long series of military successes, or diligent and skillful labours, it is generally found that the more intelligent among the Artisan and Soldier classes manifest a slight increase of their third side or base, and a shrinkage of the other two sides." This increase of angle happens at the rate of about half a degree every generation, as long as everyone continues to behave himself. As the apex angle increases, the isosceles triangle becomes closer and closer to being equilateral, with all the angles equal at 60° and all the sides the same length. In Square's own ancestry, progress was set back five generations after one poor man with an almost equilateral angle of 59.5° accidentally "transfixed" a polygon "through the diagonal." His sins were visited upon his sons, who were born with apex angles of just 58°.

Happily, Square's father was pronounced Equilateral. When this goal of 60° is reached, "the condition of serfdom is quitted, and the freeman enters the class of Regulars." From that point on, every son is a regular polygon with one more side than his father. (A "regular" polygon is one whose angles are all equal and whose sides are all the same length, so an equilateral triangle is a "regular triangle." A regular quadrilateral is better known as a square, and then we speak of regular pentagons, regular hexagons, and so on.) Our protagonist, A. Square, is a square because his father was an equilateral triangle. His sons are regular pentagons and his grandsons are regular hexagons. They are thus of a higher social class and therefore his superiors—in Flatland the commandment would be not "honor thy father and mother," but "honor thy sons and grandsons." It makes child-rearing somewhat challenging, I imagine. At the very apex (so to speak) of society are the polygons with so many sides that they are almost indistinguishable from, and are referred to as, circles. Nobody would be so ill-bred as to actually attempt to count the sides of these noblemen. "It is always assumed, by courtesy, that the Chief Circle has ten thousand sides."

I expect you have several questions at this point. How is it, for example, that given the increase in sides or angles of each generation, not everyone is by now a circle? The answer is that fertility seems to lessen as your social standing increases. The underclasses are constantly reproducing (but only rarely creating equilateral triangles), while circles may have one child at best. Regularity may also be disrupted by moral failings, and some children of "good" families may be born with many sides that are slightly irregular. This flaw can sometimes be corrected by expensive and painful treatment at the Circular Neo-Therapeutic Gymnasium. Satirizing a belief common in Victorian England that the poor can't help being stupid and venal because it's in their nature, Square asks, rhetorically, "Why blame the lying, thievish Isosceles when you ought rather to deplore the incurable inequality of his sides?" Should we forgive them their sins, then? Of course not. "In dealing with an Isosceles, if the rascal pleads that he cannot help stealing because of his unevenness, you reply that for that very reason, because he cannot help being a nuisance to his neighbours, you, the Magistrate, cannot help sentencing him to be [executed]—and there's an end of the matter."

Given the vital importance of social rank in Flatland, it is imperative to know, when you meet someone, what shape he is. In our three-dimensional world, which Square calls Spaceland, we have no trouble distinguishing a square from a triangle, because we can look at it from above and see its angles and count its sides. But if you exist on a flat plane, this is impossible. Every polygon looks like a line. Abbott/Square gives us a picture to illustrate the problem, comparing the view of a triangle and of a pentagon. "It will be obvious," says Square, "to every child in Spaceland who has touched the threshold of Geometrical Studies, that, if I can bring my eye so that its glance may bisect an angle (A) of the approaching stranger, my view will lie as it were evenly between his two sides that are next to me (viz. CA and AB), so that I shall contemplate the two impartially, and both will appear of the same size." These two shapes appear indistinguishable to the two-dimensional eye.

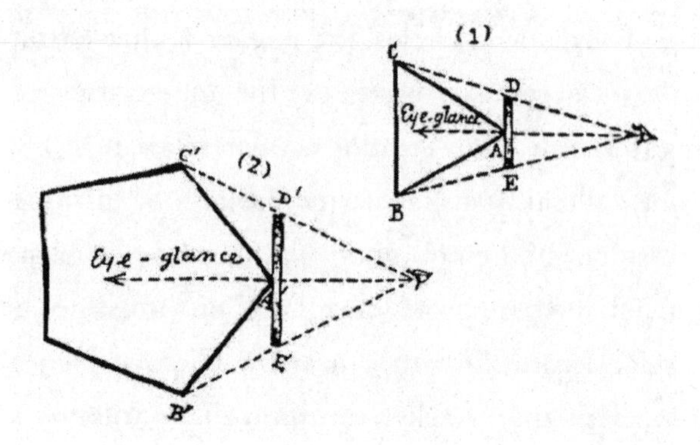

Fortunately, the atmosphere in Flatland is slightly foggy. This means that objects in the distance are dimmer than objects nearby, and we can tell triangle and pentagon apart because the triangle's edges recede more rapidly into the distance than those of the pentagon. Through years of careful training, one can learn to decipher these slight gradations of light, and thus avoid the shame of accidentally addressing a pentagon as though he were a triangle. Only triangles and squares can appear to be women (the horror), and only then from certain angles, because all higher polygons have at least two sides visible from any viewpoint. This analysis by light assumes regularity—that all the angles of a given figure are the same. Anyone born with significant irregularity, therefore, is an existential threat to society. Imagine seeing an angle of 120° approaching and, assuming you were addressing a hexagon, inviting the gentleman into your house, only to make the appalling discovery that you had been associating with an irregular quadrilateral! Such aberrations must be destroyed at birth.

Square also mentions that color has been banned since an unfortunate incident in which a low Isosceles managed to paint himself in such a way that he was mistaken for a Dodecagon—a twelve-sided polygon—and managed to inveigle his way into the affections of a nobleman's daughter. It was only after their marriage that the deception was detected—of course the girl had no recourse but suicide. Part I of *Flatland* concludes with a discussion of how women are treated in Flatland, with Square arguing against their exclusion from education. In the final sentence, he makes a "humble appeal to the highest Authorities to reconsider the regulations of Female education," a plea that Abbott himself made in Spaceland on many occasions.

Part II of the book kicks into a higher gear, geometrically speaking. Square is visited by a stranger named Sphere who takes him on a spiritual journey of enlightenment in which he discovers that there are worlds with more than two dimensions. Abbott wants to show us Spacelanders that we are as stuck in our ignorance about four dimensions as Square is about three.

Before Sphere makes his entrance, Square dreams of Lineland, a one-dimensional world consisting of a single line. Men here are line segments, and women are points. Since it's impossible to pass one another within the confines of a line, the creatures of this world spend their whole lives next to the same neighbors. They see only points, so the social hierarchy is determined by length. The Monarch of Lineland is the longest line, at 6.457 inches. Since everyone's relative position in Lineland is unchangeable, reproduction certainly cannot involve proximity. It is done, apparently, by song. A line having two endpoints, the natural order in that world is that each man has two wives; when the group reproduces, one wife has twin girls, the other has a single boy. "How else," says the Monarch, "could the balance of the Sexes be maintained, if two girls were not born for every boy? Would you ignore the very alphabet of nature?"

Square tries to explain to the ignorant Monarch that there is a second dimension—left and right as well as north and south. He first demonstrates this by saying that he can see who all of the Monarch's neighbors are, and describing them. This having failed to convince, Square then "walks through" Lineland; that is, he passes his body across the line. From the Monarch's point of view, the disconcerting effect is of a line popping in and out of existence. In another dream, Square goes to Pointland, which consists of a single point. The King of Pointland is himself the entire universe that he rules. He cannot even conceive of the existence of other beings, so it is impossible to converse with him—he believes that Square's voice must simply be another aspect of his own thoughts.

We are introduced to Sphere on an evening when Square has just been chiding his Hexagon grandson for asking silly questions about geometry. Square has explained that 3^2, or 9, represents the number of square inches in a square whose side is three inches long. So the square of a number has a geometrical, not just an algebraical, meaning. The young Hexagon says that this must mean that 3^3 must also have a geometrical meaning. Nonsense, says Square. And it's at this point that Sphere appears. "The boy is not a fool," he says. There is a meaning, and a very natural one. Square is baffled as to how this stranger has appeared in his house. Because the intersection of a sphere with a plane is a circle, Square assumes he is talking to an eminent member of society and is extremely respectful. Sphere starts to explain that he is not just a circle but a "circle of circles." He passes upward through the plane of Flatland just as Square had passed through Lineland. This time, Square sees a circle growing larger and larger, then decreasing again until it completely vanishes.

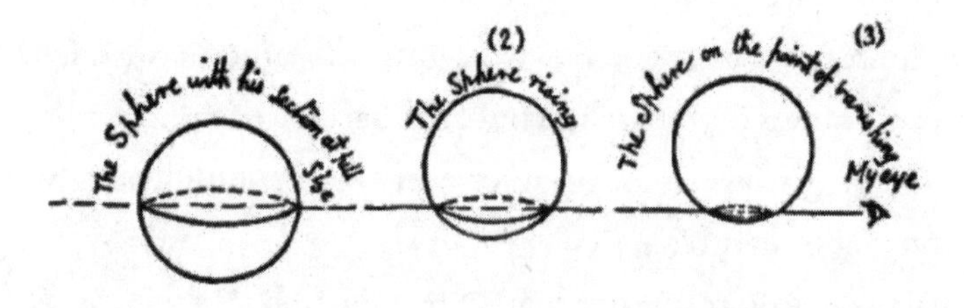

But Square cannot conceive of this "up" and "down," so Sphere resorts to analogy: "We begin with a single Point, which—being itself a Point—has only one terminal Point. One Point produces a Line with two terminal Points. One Line produces a Square with four terminal points. . . . 1, 2, 4 are evidently in Geometrical Progression. What is the next number?" Square can confidently answer that the next number is eight. Sphere replies, "Exactly. The one Square produces a *something-which-you-do-not-as-yet-know-the-name-for-but-which-we-call-a-Cube* with eight terminal Points." Sphere continues his explanation: if we define the "sides" of a shape as the boundary parts that have dimension one less than the shape, then a Point has 0 sides, a Line has 2 "sides" (the two extremal points), a Square has 4 sides, so following the pattern 0, 2, 4 . . . the "Cube" would have 6 "sides," and indeed cubes do have six square faces.

When Square still fails to grasp what is going on, Sphere finally takes him outside Flatland, to see it from above. This is, at last, what converts Square to the "Gospel of Three Dimensions." Having had his mind opened up, he asks Sphere to take him to the Land of Four Dimensions. Just as Sphere can see the inside of Square, Square asks to be able to go to the fourth dimension and see the inside of Sphere. Surely there is a fourth dimension, where, following the patterns earlier, we can extend a cube to form a shape with 16 terminal points and 8 sides? (Nowadays, the four-dimensional cube that Square imagines is typically called a *hypercube*, or sometimes a *tesseract*—the derivation may be from the Latin *tessera*, "cube," because it is a shape made of cubes.[1]) And why not a fifth, sixth, seventh, eighth dimension? Sphere, though, absolutely rejects such heretical ideas. He returns Square to Flatland, where he is imprisoned for preaching the doctrine of higher dimensions. Just as the inhabitants of Pointland, Lineland, and Flatland have been hubristic enough to assume that their worlds constitute the whole universe, the lesson for us is that we should not be, like Sphere, similarly arrogant. We should embrace the idea of the fourth dimension and beyond.

Why would someone like Edwin Abbott write a book explaining the fourth dimension? I have said that it was all the rage in the late 1800s. To understand why, we need a brief diversion into mathematical history.

The word "geometry" comes from *geo*, "earth," and *metros*, "measurement." If you want to know the size of a field, or to divide up a parcel of land into four equal parts to settle an inheritance, you need geometry, and in particular you need the geometry of the plane, because even though the surface of the earth is curved, that curvature over small distances is so slight as to be irrelevant in calculations. Surveying techniques like triangulation, in which you have two angles and one side of a triangle and can deduce the other angle and sides, can use this "Euclidean" geometry very effectively. Later, the geometry of three dimensions came along, and for astronomical uses, the geometry of spheres was needed. But the concept of a fourth dimension was never discussed.

As can sometimes happen in mathematics, one of the things that made the breakthrough possible was what seemed like a fairly small innovation in notation. Until much later than you might think, what we call algebra was all written out in words. You might say something like "The square, added to four times the number, is equal to twelve. Find the number." Nowadays we would write this as the equation $x^2 + 4x = 12$, and we could solve it using the quadratic formula or by factorizing. Amazingly, we can actually pinpoint the very first written equation, because by definition an equation is of the form "something = something else," and the equals sign was invented by Robert Recorde, a Welshman working in Tudor England. The equals sign was one of many notational inventions of his, and he decided to represent it as a pair of parallel lines because "noe two things can be more equall." The world's first equation, by the way, appearing in Recorde's 1557 book, *The Whetstone of Witte*, is $14x + 15 = 71$. Can you solve it?

For a long time, different letters were used for a thing, its square, and its cube. An expression we would write as $x^2 + 4x$ might be written instead as Q + 4N (where N is the number and Q is its square). The notation just didn't allow for x, x^2, x^3 to be "naturally" extended to x^4, x^5, and so on, much less the even more general x^n. It was Descartes who introduced the exponential notation we use today, in his book *La Géométrie* (1637), which made beautiful links between geometry and algebra. The book also established our modern convention of using letters from the end of the alphabet, like x, y, and z, for variables, and letters from the beginning of the alphabet, like a, b, and c, for constants. So if you ever wondered why we mathematicians are so x-obsessed, blame Descartes.

Meanwhile, other kinds of geometry than our Euclidean geometry of the plane were coming ever closer to being discovered. To paraphrase Hemingway's description of bankruptcy, it happened gradually, then suddenly. As we discussed in Chapter 3, concerted attempts to prove the famous parallel postulate (that given a line, and a point not on that line, there is exactly one line through that point parallel to the given line) had failed, and eventually people realized that it was completely independent of the other rules of Euclidean geometry (such as the axiom that every pair of points can be joined by a line). During the nineteenth century, mathematicians found that in fact there are geometries in which the parallel postulate does not hold true. This discovery opened a Pandora's box of new ideas in mathematics.[2]

At the same time, physicists were starting to investigate things like electricity and discovering the existence of electromagnetic fields, in which every point in three-dimensional space has not only its three spatial coordinates but additional information, or coordinates, such as the magnitude and direction of the field. This means that each point may have four, five, six, or even more numbers associated with it, and the accompanying mathematics treats these numbers in the same way as the "real" spatial dimensions. Going into higher dimensions, especially with the increasing sophistication of algebra, became the next big thing. Nowadays, we are quite comfortable with phrases like "multidimensional analysis," which really just means many numbers associated with each data point. The "dimensions" here are simply the different quantities we are measuring. For instance, in a mathematical model of the earth's climate, each point in the atmosphere will have its three spatial coordinates plus data like temperature, pressure, and wind speed and direction—that's seven dimensions already.

Pure mathematicians don't really mind if a thing doesn't "exist"—the idea of a seventy-four-dimensional hyperpyramid is just interesting in its own right. What would that mean? What even is such a pyramid? I bet it must be a seventy-three-dimensional hypercube all of whose vertices are joined to a single additional vertex in the seventy-fourth dimension. And then I want to work out how many total vertices there would be and how many edges and faces and hypercubes, and a general formula for an *n*-dimensional hyperpyramid, and so on. If you ever find a stray mathematician in the wild and decide to adopt them, just be sure to give them a lot of paper and pencils, and they will be quite happy. However, even though "being useful" and "existing" are not top priorities for the beautiful imagined worlds of pure mathematics, I will say that not long after *Flatland*'s publication came the formulation of four-dimensional space-time, of which time is the fourth dimension, along with the three standard dimensions of space. This is the perfect framework for Einstein's theory of relativity. Much more recently, physicists have been postulating an even higher-dimensional universe. If the string theorists are to be believed, the universe may in fact be ten-dimensional, or even twenty-three-dimensional. So all this fun mathematics does have proper scientific uses, if you like that sort of thing.

In *Flatland*, the fourth dimension is conceived by A. Square as being another spatial dimension.[3] That's also the interpretation used by authors positing it as an explanation for ghosts and other supernatural things, an idea ridiculed by Oscar Wilde in his 1887 haunted house parody, "The Canterville Ghost": "There was evidently no time to be lost, so, hastily adopting the Fourth Dimension of Space as a means of escape, [the ghost] vanished through the wainscoting, and the house became quite quiet." We three-dimensional beings can move about at will over a plane, stepping over the lines forming the walls of any building and appearing to change shape by altering which part of ourselves intersects with the plane. Sphere could break into any safe in Flatland and steal the contents, and a four-dimensional being could perform similar feats in our three-dimensional world. It's been proved, for example, that every knot, however complicated, can be unraveled in four dimensions. I don't know how hyperbeings manage for shoelaces, poor things.

Several authors have explored these ideas. In the 1928 short story "The Appendix and the Spectacles" by Miles J. Breuer, a Dr. Bookstrom sets himself up as a surgeon who can perform operations without the use of a scalpel, indeed without making any incisions at all. It turns out that he is a Ph.D. in mathematics, not a medical doctor. His studies of the fourth dimension, which is described as being at right angles to the usual three dimensions, have allowed him to develop a way to move the patient "along the fourth dimension," and then (for example) remove their appendix without making any cut at all.

A more chilling depiction of four-dimensional beings is given in Ford Madox Ford and Joseph Conrad's 1901 novel, *The Inheritors*. It begins when the narrator, Arthur, meets a woman who claims to come from the fourth dimension, "an inhabited plane—invisible to our eyes, but omnipresent." He is initially dismissive, but he comes to believe, or half believe at least:

> *I heard the Dimensionists described: a race clear-sighted, eminently practical, incredible; with no ideals, prejudices, or remorse; with no feeling for art and no reverence for life; free from any ethical tradition; callous to pain, weakness, suffering and death. . . . The Dimensionists were to come in swarms, to materialise, to devour like locusts, to be all the more irresistible because indistinguishable. There would be no fighting, no killing; we—our whole social system—would break as a beam snaps, because we were worm-eaten with altruism and ethics.*

The Dimensionists do indeed take over. They are inexorable. That Arthur elsewhere categorizes these cold, callous, amoral people as part of a "mathematical monstrosity" is symptomatic of a particular kind of anti-math sentiment: that numbers and equations are antithetical to all the things that make life worth living: love, joy, kindness, art. Mathematicians, according to this creed, are mere calculating machines for whom human emotions are a tedious distraction. I refute that proposition, naturally. This book forms part of the case for the defense.

. . .

In the stories we've looked at so far, the fourth dimension has been an extra dimension of space. But there's an alternative outlook. In *À la recherche du temps perdu*, Marcel Proust writes that a particular church is "for me something entirely different from the rest of the town; an edifice occupying, so to speak, a four-dimensional space—the name of the fourth being time." We are all moving along the axis of time at a rate of one second per second. If you could work out how to change this rate, then you would have invented a time machine.

Many novels feature time travel, but the grandfather of them all is H. G. Wells's *The Time Machine*. He had explored the idea of time and the fourth dimension in short stories like "The Chronic Argonauts," but it was with *The Time Machine* that the idea really took flight. The Time Traveller (as the character in that novel is known) explains matters to his friends by analogy: "You know of course that a mathematical line, a line of thickness nil, has no real existence. They taught you that? Neither has a mathematical plane. These things are mere abstractions." By the same token, he says, a cube having only length, breadth, and thickness cannot have a real existence—an "instantaneous cube" cannot really exist. "Clearly," he continues, "any real body must have extension in four directions: it must have Length, Breadth, Thickness, and—Duration."

In this interpretation, we are all four-dimensional beings. What I see of you at any moment is what you might call a temporal cross section. I see you at a particular location and at a particular time. As the Time Traveller puts it, "Here is a portrait of a man at eight years old, another at fifteen, another at seventeen, another at twenty-three, and so on. All these are evidently sections, as it were, Three-Dimensional representations of his Four-Dimensioned being, which is a fixed and unalterable thing." What the Time Traveller has done is to build a machine to allow him to move freely through time, just as we have built machines to allow us to escape gravity and move freely in the vertical direction of space.

For an entirely different account of space-time, let's meet Billy Pilgrim. In Kurt Vonnegut's novel *Slaughterhouse-Five*, Billy becomes "unstuck in time" during the Second World War. He moves constantly between different parts of his own life—he has seen his birth and death many times, and all the events between. Into this temporal chaos comes an alien race called the Tralfamadorians, who abduct Billy on the night of his daughter's wedding. They can see in four dimensions and try to help Billy understand what is happening to him. They have a very fatalistic attitude to life and death because, on Tralfamadore,

> *When a person dies he only* appears *to die. He is still very much alive in the past, so it is silly for people to cry at his funeral. All moments, past, present, and future, always have existed, always will exist. The Tralfamadorians can look at all the different moments just the way we can look at a stretch of the Rocky Mountains, for instance.*

Billy adopts their fatalism as a coping mechanism—when he hears that someone is dead, he simply shrugs and says what Tralfamadorians say about dead people: "So it goes." Our loved ones still exist, they are just elsewhen.

For such an unassuming little book, *Flatland* has cast a long shadow. Its ideas have been explored by several authors. In 1957, Dionys Burger wrote a sequel, *Sphereland*, in which a surveyor discovers a triangle whose angles sum to more than 180°. Working with Square's grandson Hexagon, who is now a trained mathematician, they realize the implication: Flatland is not a plane. They are in fact living on the surface of a very large sphere.[4] Naturally this does not go down well with the hidebound establishment.

Flatland does not concern itself with the practicalities of two-dimensional life. Obviously it's impossible for two-dimensional beings to exist . . . or is it? In 1984, A. K. Dewdney's *The Planiverse* attempted to answer just that question. The Planiverse of his conception is a two-dimensional universe, not a flat plane within a three-dimensional world. Such a universe would have to have different physical laws, and the book does an amazing job of working out some of the implications. The setup for the book is that Dewdney and a group of students create a computer simulation, 2DWORLD, of a putative two-dimensional universe. But one day, for reasons that are obscure, they start to be able to see another world that they didn't create, and to communicate with a being called Yendred, who lives on a planet called Arde. The Ardean world is circular, and the beings live on its surface—their two dimensions are east/west and up/down. The book is presented as if it is recounting real events, but there are lots of jokey references to give the game away. Among these are the fact that "Yendred" is suspiciously like "Dewdney" backward, and that one of the students in the research project is named Alice Little, which is surely a reference to the Alice Liddell for whom *Alice's Adventures in Wonderland* was written. (In the 2001 reissue of the book, Dewdney claims that a lot of people thought the original account was real, but this claim itself may have been a joke.)

Once you start thinking about how a two-dimensional civilization could work, the problems quickly mount. If you try to build a house on Arde, nobody can walk around it, because they are trapped in their two dimensions of up/down and east/west. So all buildings are underground, with "swing staircases"—staircases that lift up and down to let people move past doorways. A supporting wall cannot have a doorway in it, because every time that door was opened the house would collapse. And how is a building to be constructed? "Nails are useless, since they part any piece of material they are driven through. Saws are impossible. A beam could only be cut with something like a hammer and chisel." The solution is to construct buildings mainly with very strong glue. Meanwhile, the basic biological functions are challenging to imagine. A digestive tract that ran through the body would break the body into two parts. Tubes are impossible, and Ardeans would have to have exoskeletons, because an internal skeleton would fill the body with impassable barriers, so preventing the flow of bodily fluids. The solution for the passage of those fluids is "zipper organs" that open and close to allow bubbles of material to pass through the body. I am awed by the ingenuity of *The Planiverse*—if you like getting into technical details, it's definitely recommended. Dewdney even includes, in an appendix to the book, explanations of how the Ardeans could create machines like steam engines and internal combustion engines. It's quite astonishing what is possible.

The Planiverse opens up new realms of invention because it doesn't dismiss from the outset the idea that a two-dimensional universe can exist. That same outlook is fundamental to the most creative mathematical thinking. We've seen how in the nineteenth century we learned, with the help of *Flatland* and other books, to love the fourth dimension. Nowadays we are quite happy talking about any number of dimensions: one, two, three, four, five . . . you can go as high as you like. But imagine that Square's grandson had asked if you could have dimensions between these numbers. Of course not, he would have replied—the very suggestion of something one-and-a-half-dimensional is preposterous. And nineteenth-century Spacelanders would have agreed. But in the twentieth century a new idea blew these preconceptions out of the water. It was an idea that, toward the end of the century, gained a huge amount of traction in the public imagination. And it appears in one of the most popular novels of the 1990s. That's where we'll begin the story.

Michael Crichton's *Jurassic Park* is the story of a reckless biotech company that manages to genetically engineer dinosaurs using DNA extracted from the blood of prehistoric mosquitoes that have been trapped in amber. Being the bad guys, they decide that the way to use this amazing discovery is not to increase scientific understanding of these marvelous creatures but rather to open a dinosaur theme park, on a small island off the coast of Costa Rica. Naturally, they are certain that nothing can go wrong. Readers of a nervous disposition should look away now, because it's my solemn duty to inform you that there is a chance that if you visit, you will get eaten by a velociraptor. The park's owners arrogantly believe, because they have designed it and set it up, that they have full control over everything that happens on the island. Seemingly small mishaps or unexpected events are treated as just that—minor glitches to deal with and move on from. But nature is not a piece of clockwork. Tiny changes can become magnified until the entire system becomes unpredictably chaotic.

Crichton chooses two mathematical ways to emphasize this theme of his novel. First, there is a character, Dr. Ian Malcolm, who is an expert in chaos theory. He and two paleontologists are invited to the island as consultants. Malcolm describes how even small fluctuations in a system can lead to huge unpredictable events down the line. This is the idea encapsulated by the famous "butterfly effect." When predicting the weather, tiny changes (like the minuscule effect on air currents of a butterfly flapping its wings) have cumulative effects that can eventually mean the difference between clear skies and a hurricane. Weather systems are extremely complex, and computer simulations that predict the weather are rarely accurate beyond a few days out. The reason is that, however precisely you input your initial data (temperature, wind speeds, and so on), there will always be a tiny difference between the input and the exact real information. You might input 4.56 when the real measurement is 4.56112 . . . You cannot enter infinitely many digits, so what you put into the algorithm is always rounded. Absolute precision is impossible.

For some mathematical models, this doesn't matter. For instance, suppose you want to know where an object will end up 24 hours from now, based on your measurement of its initial position. If you know it is moving at 100 miles per hour, and you are off in its initial position by 1 mile, then the object will move 2,400 miles, and you'll still be off by 1 mile in your prediction of its final position. However much time passes, your prediction will remain just 1 mile out. In some sense, you have control over the error. But suppose now that what you are measuring is not its initial position but its initial speed, and you want to know where it will be 24 hours from now. If you get the speed out by 1 mile per hour, so that in fact the object is moving at 101 miles per hour, not 100, then each hour will increase the inaccuracy of the prediction so that after 24 hours your prediction of its position will be not 1 mile out, but 24 miles. That's a 1 percent error, and the error will double every day.

It's even worse if the error is with the acceleration. If you make an error of just 1 mile per hour in your measurement of the object's acceleration, so that instead of moving at a constant speed of 100 miles per hour its speed is slowly increasing by 1 mile per hour, then guess how far out your prediction of its position will be, just one day later, even if you get the starting position and speed exactly right? It's 288 miles, well over a 10 percent error in just one day. This same error over the course of a week would lead you to be off by 14,112 miles. Which is pretty bad, given that the distance from the North Pole to the South Pole is only 12,440 miles. The object could be more or less anywhere on Earth's surface at this point. Small initial discrepancies can spiral out of control, which is why when navigating we perform regular course corrections.

In *Jurassic Park*, the mathematician Dr. Malcolm is there to explain this issue in words, but the book also gives us a visual clue to what's happening. At the start of each section, a curious design appears, which changes and grows as the story develops, over seven "iterations." The "first iteration" looks like this:

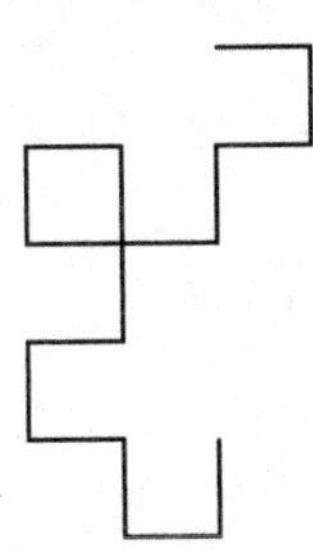

It's a fairly simple design made up of straight lines at right angles. Here's the second iteration:

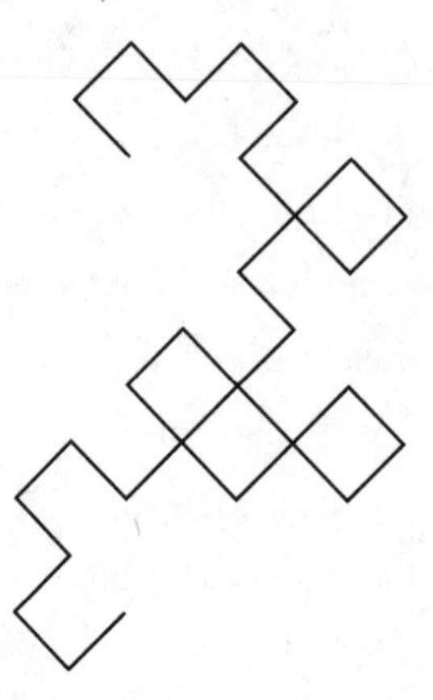

The shape that is being built up has a very simple construction rule, and you can try it yourself. Go and get a piece of paper and cut a long thin strip off it. Now bend that strip in half and unfold it, setting it on its edge, so the bend makes a right angle. You'll see that the fold you created has turned the straight strip into an L shape.

This is the first step. Now do it again. Fold it back up, then fold it in half again, then unfold. It's a slightly more complicated pattern now, but still made of straight lines and right-angle bends.

If you keep going with this process, here's the outcome after three, four, and five folds:

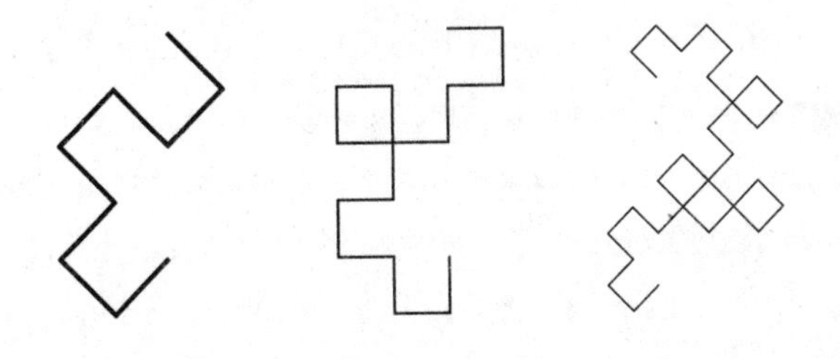

If you look at the middle picture, you'll see that the patterns of left bends and right bends are precisely those of the "first iteration." And the picture on the right, which comes from five folds, is the "second iteration." Continuing with this very simple folding game leads to the developing shape. It doesn't take long before the pictures become very complicated, and it gets increasingly difficult to guess whether, when the paper is unfolded, the next bend in the paper will be to the left or to the right. By the final section of *Jurassic Park*, when things on the island have reached a breaking point, what started as a simple design has evolved into a fearsomely complicated diagram. We could in principle carry on the process indefinitely, with the eventual outcome looking like an intricate curved shape, some parts resembling other parts at different scales. Here are the third, fourth, and fifth "iterations":

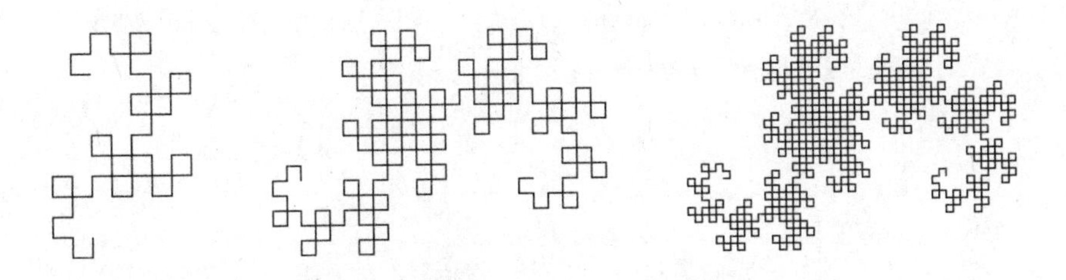

There has been a teeny bit of cheating here on the part of the illustrator of *Jurassic Park*, because a few steps have been missed. We already jumped straight to four folds to get the first iteration. That's fair enough, because the first few steps really aren't very promising. The second iteration corresponds to five folds, and the third to six folds. But in programming the next few folds, I realized that the fourth iteration, which should correspond to seven folds, actually corresponds to eight. The fifth iteration arises from ten folds, the sixth is from twelve, and the seventh is probably from fourteen, though by the time we get to that stage, it's hard to detect the individual lines because the resolution is not good enough. Here are my versions of the sixth and seventh iterations:

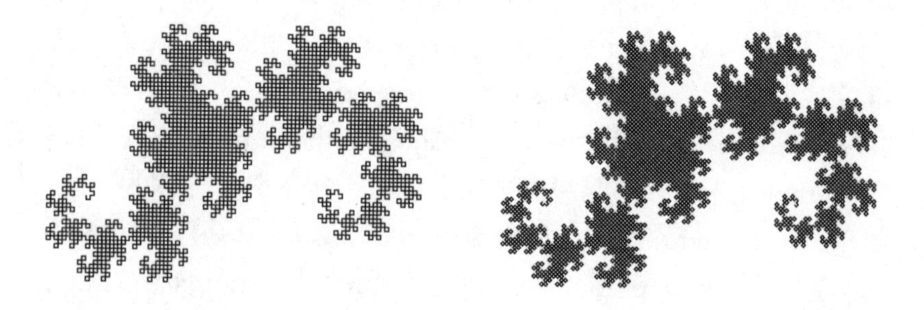

This shape was discovered by NASA physicist John Heighway and christened the Heighway dragon by his colleagues William Harter and Bruce Banks, who, together with Heighway, were the first to explore its properties. Nowadays it's usually simply called the *dragon curve*. It entered the popular consciousness in 1967, when it appeared in Martin Gardner's Mathematical Games column in *Scientific American*, illustrated with a diagram something like this one:

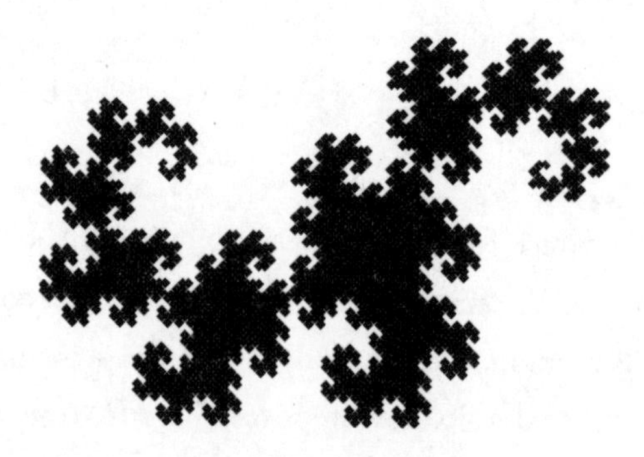

He wrote, "The curve vaguely resembles a sea dragon paddling to the left with clawed feet, his curved snout and coiled tail just above an imaginary waterline." Rather unfortunately, the resemblance to dragons is weaker in the versions shown in *Jurassic Park*, because they are shown upside down compared to the diagram given in *Scientific American*. They are, as Harter joked, dead dragons.

The dragon curve is an example of what's known as a *fractal*—a shape that is produced by a repeated, or iterative, process continued indefinitely. Like the number π, we can never produce them perfectly, because we cannot in real life complete infinitely many steps. It's hard, by hand, to carry out even a few iterations. I didn't produce the dragon curve pictures by folding paper. There's an alternative way to do it, which is a simple repeated process that can be explained to a computer. At each stage, you replace every straight line by a bent line, alternating bending left and right. You can see what I mean if I show you it happening—this is step 3 moving to step 4, and I've kept the lines from step 3 as dotted lines—each of them has been replaced by a bent line. I've shown an arrow in the first step, starting at the top, where we replace the first line with a left bend:

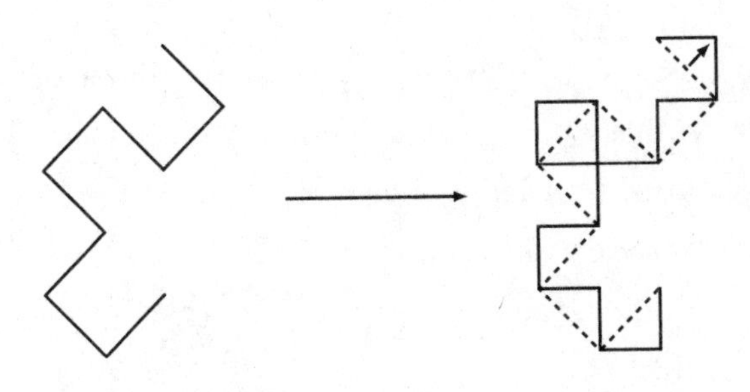

Michael Crichton found the dragon curve to be a compelling illustration of the fact that huge and unpredictable complexity can emerge from even the simplest starting points—and it gets bonus points for its appropriately reptilian name. In *Jurassic Park*, the exploration of ideas around chaos and complexity is interesting and worthwhile. Is this literature? I say yes. Don't get me wrong, some of the writing is rather "He pulled up the schematics from the mainframe and punched the coordinates into the console," but it's still a damn good read.

The use of fractals in a book like *Jurassic Park* is an indication of their reach in the popular culture of the time. At the turn of the 1990s, when the book appeared, fractals were having quite a moment. Fractal art, inspired by the famous Mandelbrot set (named for the mathematician Benoit Mandelbrot, who coined the term "fractal"), was appearing on dorm room posters, magazine covers, and T-shirts. Literature caught on to the trend too. John Updike's 1986 novel, *Roger's Version*, features a computer scientist, Dale, repeatedly zooming in on computer-generated fractal structures to seek the hand of God. Tom Stoppard's wonderful 1993 play, *Arcadia*, incorporates fractals along with several other mathematical ideas—we'll have more to say about it in Chapter 10. Why did fractals suddenly take off like this? The answer is to do with the way they are constructed.

It's hard to imagine a simpler process than folding a strip of paper, or a more basic iteration than "replace each straight line with a bent line." If you try to draw the dragon curve by hand, it's quite frustrating because you have to erase previous steps to add in all the bent lines. But there is a fractal that I used to enjoy doodling during chemistry classes (sorry, Dr. Vuik) with the advantage that you don't need to delete previous steps. You start with a simple triangle. Then, at each stage, you add a triangle to the middle third of each straight line. The first three steps look like this:

After infinitely many steps (or at least my approximation of infinity, which happens to be six) we get this appealing shape, which is called the *Koch snowflake curve*:

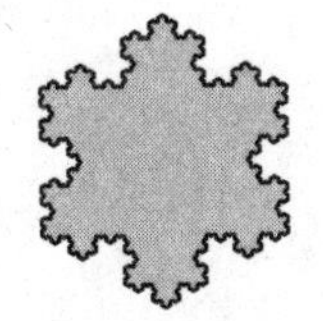

The Koch curve is one of the oldest fractals to be discovered—it was described by the Swedish mathematician Helge von Koch in a paper way back in 1904, long before the word "fractal" was invented (it gets name-checked in *Roger's Version*, incidentally). It's an unusual fractal in that the first few hand-drawn steps can give us a good idea of the eventual shape. But for fractals like the dragon curve and others, we can really start to make sense of things only after many steps. That's why the study of fractals didn't really take off until we could start using computers to calculate hundreds or even thousands of iterations, and that's why they exploded onto the cultural scene in the late twentieth century.

Where does this word "fractal" come from? Well, think of a square of side length 1 (1 centimeter or 1 inch, or any other unit, take your pick). If we multiply the length of the side by 3, we get a square of area 9. In general, if we multiply the length by x, we multiply the area by x^2. The 2 in x^2 comes from the fact that squares are two-dimensional. For a three-dimensional shape, like a cube of side length 1, trebling the length of each side gives a cube of volume 27. In general, if we multiply the length by x, we multiply the volume by x^3. So the dimension is 3. (This may remind you of our discussion of the square-cube law earlier.) In the simplest case, multiplying the length of a line by x gives a line of length x^1, confirming the fact that lines are one-dimensional. Now, what is the dimension of the Koch curve? It is easier if we work with just one side of the triangle, so we start with a line of length 1, then replace the middle third of the line with two line segments equal in length to the portion removed, making that little extra triangle, and so on. What happens to the curve if our starting line is trebled in length? The picture below shows the starting line and the finished curve in each case:

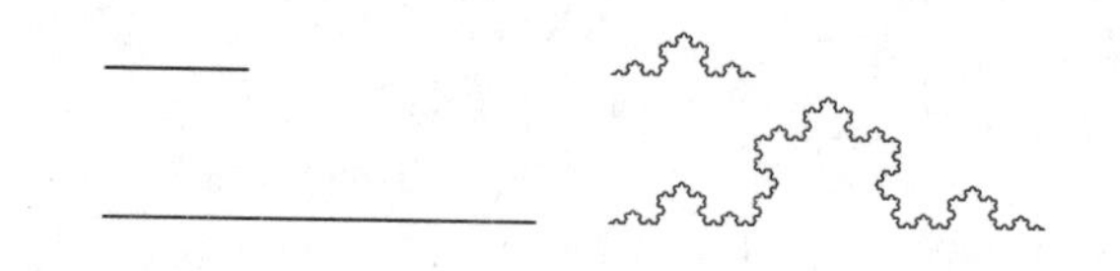

In the finished curve we now have four copies of the curve that was made with the original line. So multiplying the line length by 3 had the effect of multiplying the curve's length by 4. Since 4 is more than 3^1 and less than 3^2, the dimension of the Koch curve lies between 1 and 2. It is some fractional quantity, not a whole number. It turns out that the dimension is roughly 1.26 (because $3^{1.26} \approx 4$). Hence the word "fractal": they have *fractional dimension.*

Finding the dimension of the dragon curve is a little more involved, but it is roughly one-and-a-half-dimensional. So we can, after all, have dimensions between the whole numbers, an idea that would have seemed as strange to Edwin Abbott Abbott as the idea of three dimensions was to A. Square.

That's all very well, you say. But surely there's no such thing as a (−1)-dimensional space, right? The lessons of literature, and of mathematics, tell us we would be unwise to rule it out.

Although the Koch snowflake was discovered at the very start of the twentieth century, a deeper exploration of fractal geometry really became possible, as we've seen, only when the technology caught up. Whole new vistas of geometry were opened up by the invention of computers. Something exactly analogous happened in the field I want to spend the rest of this chapter exploring: cryptography. The history of code making and code breaking is long, and everyone loves a secret code, so it's no surprise that they crop up in fiction. But we have to wait until 1843 for a story that really revolves around cryptography. That was the year that Edgar Allan Poe won a $100 prize for "The Gold-Bug," an enjoyable code-cracking adventure in which the notorious pirate Captain Kidd leaves encrypted instructions for finding a fabulous hoard of treasure.

Why so late? After all, secret messages have been exchanged for thousands of years. One instance, dating back to 499 BCE, is recounted by the ancient Greek historian Herodotus. It seems that Histiaeus, ruler of the city of Miletus, wanted to send a secret message to his ally Aristagorus instigating a revolt against the Persians. So he shaved the hair of a trustworthy slave, tattooed the request onto his head, and waited for the hair to grow back. Then the slave was sent off to Aristagorus, who could shave the slave's head and read the message.

Methods like this, whereby you hide the message rather than put it in code, are called *steganography*, from the Greek for "hidden writing." The problem is that if this is all you do, then should the message be discovered, all is lost—assuming the discoverer can read. Until the comparatively recent past, most people were illiterate, so this was less of an issue. Data from the United States are available only from about 1870, but in the United Kingdom, literacy was less than 20 percent until the sixteenth century; by 1820 this had jumped to nearer 60 percent (this is unrepresentative of the global picture, though; it's estimated that worldwide at this time only 12 percent of people could read or write). When most people can read, suddenly steganography starts to look insufficient. You have to start encrypting your secret messages in other ways—and that's where cryptography comes in. That tipping point was likely reached in America around 1800.

Poe himself had a long-standing interest in cryptography—perhaps not surprising when you consider that a secret code had been, two years before Poe's birth, a crucial part of one of the most sensational trials of the century: that of Aaron Burr for treason. Burr had sent an encrypted message appearing to state his intention to establish an independent country in parts of some southern states and Mexico. The recipient of the message, General Wilkinson, decrypted it and then sent it to the president, Thomas Jefferson. But during the trial it emerged that Wilkinson had doctored the message to make himself look innocent. Ultimately, Burr was acquitted.

Public interest in cryptography was high enough in 1839 that when Poe challenged readers of a Philadelphia magazine to send in their encrypted messages, he received hundreds of letters. (He later claimed that all the ciphers had been easily solved, except one that was an obvious hoax.) He went on to publish a series of articles on "Secret Writing" in 1841, during his time as editor of *Graham's Magazine*. As he points out in these articles, the exchange of secret messages has been taking place for thousands of years, using a multitude of techniques. He used two of these techniques in "The Gold-Bug."

Poe was a master storyteller. His gothic tales ("The Tell-Tale Heart," "The Fall of the House of Usher") still make us shudder today, and he has a good claim to have invented detective fiction with "The Murders in the Rue Morgue." He was also a successful poet—"The Raven" brought him instant fame—as well as a brilliant magazine editor and literary critic. It has to be said that Poe did not pull his punches as a critic. The poet James Russell Lowell once joked that he "seems sometimes to mistake his phial of prussic-acid for his inkstand." Comparing two writers, Cornelius Mathews and William Ellery Channing, Poe said that "if the former gentleman be not the very worst poet that ever existed on the face of the earth, it is only because he is not quite so bad as the latter. To speak algebraically:—Mr. M. is e*x*-ecrable, but Mr. C. is x plus 1-ecrable."

This pun hints tantalizingly at a mathematical inclination, and there's further evidence in Poe's writing. In an essay on poetry, he writes that the subject is very amenable to analysis: "One tenth of it, possibly, may be called ethical; nine tenths, however, appertain to the mathematics." In a story about a hot-air balloon flight,[5] the protagonist, Hans Pfaall, is able to calculate altitude by a "simple" application of spherical geometry, remembering that "the convex surface of any segment of a sphere is, to the entire surface of the sphere itself, as the versed sine of the segment to the diameter of the sphere." Unsurprisingly, Poe scored very highly in mathematics when attending the US Military Academy; a fellow student said he had a "wonderful aptitude."

While Poe remarked on the undoubted power of mathematics and its importance in training the mind to think analytically, he was careful to stress that mathematical skill in the abstract was not enough: a true genius must be able to make logical inferences in the real world. There's an interesting conversation in his story "The Purloined Letter" between the protodetective Auguste Dupin and the narrator in which Dupin is explaining why a certain suspect, the Minister, has been underestimated by the Prefect of Police. The Prefect believes that all fools are poets and thus deduces erroneously that all poets (including the Minister) must be fools. The narrator counters that the Minister "has written learnedly on the Differential Calculus. He is a mathematician, and no poet." No, says Dupin, he is both. "As poet *and* mathematician, he would reason well; as mere mathematician, he could not have reasoned at all." This isn't quite how I might phrase it—but mathematicians tend to believe something very similar. A true mathematician is not a mere calculating whiz. They must also have intuition, a sense of the beautiful. For Poe's Dupin, and all the Sherlock Holmeses and Hercule Poirots who have followed him, the real magic happens when you apply the powerful techniques of inference by pure logic *outside* the confines of abstract mathematics. The meeting point of abstract analysis and real-world intuition is cryptography.

Let's see how Poe used cryptography in "The Gold-Bug." The story tells how down-on-his-luck William Legrand discovers instructions for finding treasure that are concealed using both steganography (in this case, invisible ink, which is revealed by the application of heat) and cryptography. When the message does appear, it is an encrypted sequence of symbols. Legrand guesses that a so-called *substitution cipher* has been used—in which each letter is replaced by a different symbol.

Substitution ciphers date back at least two millennia. The earliest known use of the idea dates all the way back to Roman emperor Julius Caesar. He replaced each letter by the one three farther along in the alphabet, so that *a* becomes *d*, *b* becomes *e*, and so on. This simple technique is still known as a *Caesar shift*. But it wasn't just military commanders using these techniques. The *Kama Sutra* describes, among (ahem) other things, sixty-four arts that women should study; along with singing, dancing, flower arranging, "arithmetical recreations," and poetry composition, item forty-four is "the art of understanding writing in cypher, and the writing of words in a peculiar way." One of the techniques known to be in use at the time was a kind of substitution cipher in which letters were paired together. This has the advantage that encryption and decryption involve the same process: if, say, *a* and *q* are paired, then *a* is encrypted as *q*, *q* is encrypted as *a*, and the same procedure will decipher the message.

But substitution ciphers can be much more sophisticated than this. They can use any rearrangement of letters and can also replace them with any symbols. On the face of it, deciphering such a code looks like an insuperable challenge. You can't just try all the possibilities. There are $26 \times 25 \times \cdots \times 2 \times 1 = 403{,}291{,}461{,}126{,}605{,}635{,}584{,}000{,}000$ ways to shuffle the twenty-six letters of the English alphabet. Happily, a mathematical analysis of language can help. Mathematical analysis to aid in decryption dates back at least to the ninth century CE, when the Islamic philosopher and mathematician Al-Kindi wrote his *Manuscript on Deciphering Cryptographic Messages*, which explained how to perform a frequency analysis. It's a powerful technique that is almost guaranteed to work if you have a sufficiently long sample of the encrypted text. I mentioned in Chapter 3, when we talked about lipograms, that if a text has been encrypted by substituting letters for other letters or symbols, then you can make the educated guess that the most commonly occurring symbols in the ciphertext correspond to the most commonly occurring symbols in English (or whatever language the message was written in).

This is precisely what Legrand does in "The Gold-Bug." The most frequently used letters in English are *e*, *t*, and *a*, and it would be a highly unusual text that did not have *e* as by far its most common letter. So you have a good idea which letter of the ciphertext represents *e*. It also helps to know common letter pairs and words. The words "the" and "and" are likely to occur regularly. Legrand explains that once he has guessed that *8* represents *e* on this basis, he finds the string "*;48*" occurring seven times: it's likely, then, that *;* means *t* and *4* means *h*. This is especially likely because *;* is the second most frequently seen symbol in the message and so probably represents *t* just on that basis. However, Poe/Legrand makes life harder for himself by using an incorrect frequency list: he claims that, in order, the letters appearing most regularly in English are *e*, *a*, *o*, *i*, *d*, *h*, *n*, *r*, *s*, and *t*, which massively underestimates *t* and overpromotes *d*.

Nowadays, it's very easy with computers to analyze large amounts of text and find their frequency distributions. But back in Poe's day and before, this would have been a much more challenging job. When Samuel Morse was trying to work out which characters should be quickest to send by telegraph, in order to construct the most efficient code, he came up with an ingenious shortcut. At that time, printers would manually typeset pages by arranging individual letters on the page. They would therefore have many more of the most commonly used letters on hand. Morse simply counted the number of each letter that the printer kept, to get a rough idea of the relative frequencies of each letter. This is much quicker than analyzing pages and pages of text by hand.

Are there any ciphers that aren't susceptible to this kind of analysis? One example appears in Jules Verne's 1864 novel *Journey to the Center of the Earth*. In the story, the eccentric but brilliant Professor Liedenbrock and his nephew, Axel, decipher an ancient coded parchment that gives instructions to make said journey. The encryption this time is what's called a *transposition cipher*. These ciphers are essentially a prespecified anagram of the message. The advantage here is that, with an anagram, frequency analysis tells you nothing—the number of occurrences of each letter matches exactly that of the original message, because its letters have simply been rearranged rather than substituted. Here's an example:

Suppose I want to encrypt the message "Pure mathematics is, in its way, the poetry of logical ideas" (something Einstein is supposed to have said). I first write it vertically in columns, so that to read it you work down the columns from left to right:

P	t	i	n	y	e	l	l
u	h	c	i	t	t	o	i
r	e	s	t	h	r	g	d
e	m	i	s	e	y	i	e
m	a	s	w	p	o	c	a
a	t	i	a	o	f	a	s

Then you copy down the rows as a horizontal message: "Ptinyell uhcittoi resthrgd emiseyie maswpoca atiaofas." The original message is now rather effectively obscured.

But if you know what has been done, you can retrieve it simply by reading off the first letter of each "word," then the second, and so on. (Alternatively, you can copy it into a 6×8 grid and read down the columns.) Much more complicated permutations are possible, but as long as I know the permutation that has been carried out, I can crack the code.

If I don't know it, then the task gets a bit harder. Frequency analysis is useless. However, with a column method like this, the length of the message gives you a big clue (even if I didn't helpfully leave a space between each "word"). Our message is 48 letters long. We created a 6×8 rectangle (and $6 \times 8 = 48$). There are really only a few possibilities to try, which we can get from the numbers that divide 48. If words of 8 letters don't work, we can try words with 6 letters, and failing that, we can quickly exhaust all the other possibilities: 2, 3, 4, 12, 16, 24. No point trying 1 or 48, as that gives us back the original message straightaway.

This column cipher is the basis of the secret message in *Journey to the Center of the Earth*. There are some obscuring factors—the message is written in Icelandic runes that must be converted into the Latin alphabet, and then when the message is untangled it still looks wrong because it has been written backward. But young Axel and his uncle get there pretty quickly once they hit on the correct cipher. And that's the case in many, perhaps most, precomputer cryptography stories. The toughest challenge is to determine what code has been used—and there are many. Sherlock Holmes, with his usual modesty, pronounces himself to be familiar with all forms of secret writing: he is "the author of a trifling monograph upon the subject, in which I analyse one hundred and sixty separate ciphers." The only thing to do is to try every cipher you know and see if you get anywhere.

One encryption method in fiction that surprised me was a trick, in a 1906 short story by O. Henry, that basically invents predictive text almost a century before cell phones. In "Calloway's Code," a journalist has to get a message across enemy lines, past censors who will destroy it if they find any pertinent information. He sends what appear to be garbled phrases like "brute select." Eventually, a young reporter by the name of Vesey realizes that the code is "newspaper English." All you have to do is think what word always follows the given word in the clichéd language of newspapers: "brute force," "select few." Then "brute select" becomes "force few"—in other words, the army is small. Calloway's editor is torn: on the one hand, Vesey has helped the paper get a great scoop; on the other, his method reflects rather badly on the literary standards of his newspaper. "I will let you know in a day or two," he says, "whether you are to be discharged or retained at a larger salary."

Just like the study of fractals, the biggest strides in cryptography have become possible only since the invention of computers. In fact, the connection is even stronger. One could argue that computers were invented, at least in part, to crack codes. Many books, plays, and films have told the story of the code breakers who cracked the Nazi Enigma machine cipher in World War II, the most well known of which is perhaps Robert Harris's novel *Enigma* (adapted into a 2001 movie of the same name). The German Enigma machines had multiple dials that were put into new positions every day, specified by a codebook given to operators. Even if you had access to a machine, it was useless without knowing the settings. Each day, the race to crack that day's code had to begin anew because the settings were changed.

To give you an idea of the astronomical challenge involved, let me tell you just a little about the Enigma machine. It looks a bit like a small typewriter. The operator types a message, which the machine encrypts; the encrypted message is then sent to a receiver and deciphered using another Enigma machine. First, the operator inserts three of a possible five "scramblers," in a specified order, each in one of twenty-six possible orientations. There are $5 \times 4 \times 3 = 60$ ways to choose the scramblers, and $26 \times 26 \times 26 = 17{,}576$ ways to arrange them once chosen. Already this is far more than could be checked by hand. But it gets worse. Inserted between the keyboard and the scramblers is a "plugboard" that swaps ten pairs of letters. The number of ways to choose ten pairs of letters from an alphabet of twenty-six is vast—150,738,274,937,000, or around 151 trillion. The total number of Enigma settings is the product of the 60 ways to choose the scramblers, the 17,576 ways to arrange them, and the 151 trillion plugboard settings. It is an incredible 158,962,555,218,000,000,000. Even if you could invent a machine that could check a billion settings per second, it would take more than five thousand years to work through all the possibilities. And the settings, remember, change every day. No wonder the Nazis thought it was unbreakable.

But here's where the mathematician Alan Turing comes in. He came up with a brilliant insight that allowed him to cancel out the effect of the plugboard, those 151 trillion additional combinations. Working with a team of cryptanalysts, he designed a machine called the Bombe that could work through the 17,576 scrambler possibilities for a given set of scramblers. Several Bombes would run in parallel, each working on one of the sixty choices of scramblers. Eventually the Allies were able to crack each day's code within hours, and the Germans had no idea. It's estimated that this breakthrough shortened the war by two years. Alan Turing was a gifted mathematician who tragically died before his astonishing contribution to the war effort could be made public. Hugh Whitemore's 1986 play, *Breaking the Code*, poignantly tells this story. The end of Turing's life was made miserable after he was prosecuted for homosexuality, and he died, almost certainly by suicide, after eating an apple laced with cyanide. It's often said (though sadly it's probably apocryphal) that the Apple logo is a tribute to Turing.

Now that computers are on the scene, whole new vistas of cryptography have opened up. All recently developed encryption methods rely on mathematics. Many a thriller has featured a genius cryptographer saying something like "Jeez, they're using a quantum elliptic curve encryption algorithm with a thousand-twenty-four-bit key," but that's more window dressing than mathematics. A book that really does involve modern mathematical ideas about cryptography is Neal Stephenson's *Cryptonomicon*. If you want to learn a lot more about cryptography than I have the space to tell you, and you want to do it in the form of a brilliantly exciting, funny, and suspenseful nine-hundred-page epic that features expressions like

$$\zeta(s) = \sum_{n=1}^{\infty} \frac{1}{n^s}$$

and

$$\pi = 4\sum_{n=0}^{\infty} \frac{(-1)^n}{2n+1}$$

in its very first chapter, then *Cryptonomicon* is the book for you.

For a different experience, I can't resist sharing with you a few of the delights of Dan Brown's oeuvre. I enjoyed *The Da Vinci Code* thoroughly. But, my goodness, there's a lot of mathematical nonsense in it. Here's a "mathematician" in the book talking about the golden ratio, otherwise known by the Greek letter phi (ϕ): "As we mathematicians like to say, PHI is one H of a lot cooler than PI." No. No, we don't. I mentioned the golden ratio in Chapter 2, when we were talking about the Fibonacci sequence 1, 1, 2, 3, 5, 8, 13, . . . The sequence of the ratios of consecutive terms of this sequence converges to ϕ, which equals $\frac{1}{2}(1+\sqrt{5})$. It's an interesting number, but one H of a lot of guff is spouted about it, not least by Dan Brown. No, Leonardo da Vinci's *Vitruvian Man* is not based on it. Nor did the Roman architect Vitruvius (which is where the Vitruvian part comes from) make any claims about the golden ratio and the human body. Oh, and while we are here, it's not true that "mathematician Leonardo Fibonacci created this succession of numbers in the thirteenth century." The worst bit (brace yourselves) is when the hero of the book, Robert Langdon, professional "symbologist," breaks the hearts of mathematicians everywhere when he says that phi equals 1.618. That's just an approximation of it, because, like its much cooler friend π, it can never be fully written down—it goes on forever. To cut it down in its prime like that, when part of its allure is its mysterious infinitude . . . it's a tragedy. In fact, the other name for ϕ, the Divine Proportion, was coined by the sixteenth-century Italian scholar Luca Pacioli because, like the divine, it can never be fully known. It does not "equal" 1.618. All right, I'll stop now. But there should be trigger warnings for mathematicians at the start of that chapter of *The Da Vinci Code*.

Anyway, the novel features a beautiful young female cryptographer, Sophie Neveu, and a rather older male academic, Robert Langdon, in a race against time to uncover the secret of a shocking conspiracy at the heart of the Catholic Church. The first encryption method they encounter is a simple anagram. Later on, we meet the Atbash cipher, an ancient cipher originally used in the Hebrew alphabet. Its name tells us how to use it. The Hebrew alphabet begins aleph, beth, gimel, daleth (roughly *a*, *b*, *g*, *d*), and ends qoph, resh, shin, taw (*k*, *r*, *sh*, *t*). The cipher simply reverses the alphabet. Aleph swaps with taw, beth swaps with shin, and so on. That is, $a \leftrightarrow t$, $b \leftrightarrow sh$: Atbash. In English we might call it the Azby cipher. I don't know about you, but if I were the leader of an ancient and powerful society charged with preserving the truth about the Holy Grail, I might use something a bit more secure.

Dan Brown's *Digital Fortress* is a totally different book. It features a beautiful young female cryptographer and an older male academic in a race against time to uncover the secret of . . . hang on a minute! Okay, maybe there are some similarities. This time the cryptographer[6] works for the US National Security Agency. ("Susan Fletcher's legs. Hard to imagine they support a 170 IQ" is an example of the style of writing.) She and academic David Becker (legs presumably able to support whatever his IQ is) are drawn into a convoluted scenario involving the NSA trying to prevent the release of an "uncrackable" encryption method.

A lot of big cryptography words are used in *Digital Fortress*, but they are entirely irrelevant to the story because the actual codes that Susan and David solve are at least two thousand years old. I'll mention just one. It's attributed in the book to Julius Caesar, who Susan says is the "first code-writer in history"—we'll pass over that one in silence—and it's a special case of the column cipher used in *Journey to the Center of the Earth*. The example I gave before had 48 characters, which we arranged into a 6×8 rectangle. This Caesar variant has an extra requirement: the grid used must be a square one—the same number of rows as columns. This makes life easier for the recipients, because there is no trial and error required. If you receive a message 144 characters long, you simply take the square root of 144, which is 12, copy the message into a 12×12 grid, and read down the columns to retrieve the message.

The consequence of this method might already have struck you: most numbers don't have an exact square root. No problem, apparently. According to Susan, each message has a letter count that is an exact square number. Seems a bit implausible. There's poor Caesar, sweat dripping off his toga, brow furrowed, the future of Rome at stake, trying to work out how, by Jupiter, he can express himself in a perfect square number of letters. Fortunately, there's a simple solution: Your message can be any length you like. At the end, you simply pad it out with enough letters to get up to the next square number. To send a twelve-letter message, for instance, Caesar can just add four letters at the end so that the total is 16 (4 squared), then follow the usual procedure. When the message is received, perhaps it reads vvvxeiiondcxiiio. A quick count gives sixteen letters. The square root of that is four, so we write the message into a 4 × 4 square grid:

v	v	v	x
e	i	i	o
n	d	c	x
i	i	i	o

We can then read off the original message column by column. The last four letters, unless Caesar is being unexpectedly informal, can be discarded.

I want to finish the chapter with two cryptographic techniques that use numbers. Having said that essentially all encryption algorithms since the invention of computers rely on mathematics, I thought I had better give at least one example. It's called *RSA*, these being the initials of the second set of people to invent it. The first person was a mathematician named Clifford Cocks, who was at the time working for the British equivalent of the NSA, the Government Communications Headquarters. His work was classified, so nobody knew about his discovery until many years later. The idea behind it is ingenious. The encryption method can be made completely public, encrypted text can be printed on the front page of every newspaper, and still it can't be cracked. It's based on the mathematical observation that while it's very easy to multiply numbers, it's really hard to factorize them—to break them up into their component parts.

What I mean is this: given a piece of paper, you could quite quickly find 89 × 97, but if I asked you to find all the numbers that divide exactly into 8,633, it would take ages—you'd have to just keep trying numbers until you found one that worked. (I won't keep you in suspense: 8,633 is actually 89×97, and these numbers are both prime, so the only factors are 1, 89, 97, and 8,633.) This lopsidedness in the difficulty of multiplication compared to factorization is the basis of RSA.[7] A number N that's the product of two very large primes p and q is made public, but the primes themselves are not revealed. Then messages are encrypted using this N (essentially by converting them into a number, raising that number to a large power, dividing by N, and transmitting the remainder). There's a neat mathematical trick to reverse this process and retrieve the original code, but it can be done only if you know p and q. Because there's no known quick method to factorize large numbers, p and q can't be determined by an adversary, even when they know N, so as long as you use big enough numbers, the code can't be cracked.

The other encryption method, which is much older, also uses numbers, but in a completely different way. It's called a *book cipher*, and an example of it occurs in the Sherlock Holmes story *The Valley of Fear*. The setup is simple. If you and I want to send each other secret messages, we agree in advance on a book that we both own. To send you a message, like "Cover blown," I have to find the words "cover" and "blown" somewhere in the book. If "cover" appears as the sixth word on line twelve of page 132, I send 132 12 6, while 415 3 15 would represent the fifteenth word of the third line of page 415. There are variations of this method; for instance, you can instead list words on the page rather than line, but this technique involves more counting. The code is uncrackable without knowing the book. If we were suspected, though, our enemy could in extremis try all the books that both of us own. In *The Valley of Fear*, Sherlock Holmes faces the challenge of trying to crack such a code without knowing the book. He is helped by the fact that the page number 532 is given, which means the book has at least 532 pages, and that the code gives a column number as well—how many books are printed in columns? This narrows things down enough for Holmes and Watson to find the book and crack the code.

To finish this chapter, I'm challenging you to crack a book cipher.

But what's the book? It has to be something both you and I own. How could I possibly know of a book that you are guaranteed to have in your possession at this precise moment? With that conundrum ringing in your ears, I'll finish the chapter with the code. Good luck!

26 13 1

41 11 2

137 31 3

9 17 9

15 2 7

9

The Real Life of Pi

Thematic Mathematics in the Novel

I am a person who believes in form, in harmony of order. . . . I'll tell you, that's one thing I hate about my nickname, the way that number runs on forever." So says "Pi" Patel, the narrator of Yann Martel's Booker Prize–winning novel *Life of Pi*. It's the story of a shipwrecked boy who survives by spending 227 days in a lifeboat with a Bengal tiger named Richard Parker. Pi, the famous mathematical constant relating the circumference of a circle to its diameter, is a fascinating number, and, as Pi Patel says, it does indeed go on forever. It is "irrational"—it cannot be written as a fraction or a decimal that terminates. This idea of "the irrationality of Pi" is also, in a play on words, a key theme of the novel—we cannot ever be sure how much, if any, of his dreamlike experience was real and how much imagined.

In this chapter, I'll show you some of the ways that elemental mathematical ideas have been used to illuminate or advance the themes of a narrative. In Chapter 8 we saw how literature has responded to the fads and fashions of popular mathematics. Now we'll see how authors have engaged with timeless mathematical themes: the properties of numbers like π, ideas of the infinite, and even the nature of mathematical thought itself.

In Yann Martel's novel, Pi Patel recounts his early life in Pondicherry,

India. He tells us that he was named after a swimming pool—the Piscine Molitor in Paris—because a close family friend and champion swimmer used to talk so much about visiting it during his youth. Unfortunately, "Piscine Patel" sounds too much like "Pissing Patel," and after years of teasing he decides drastic action is required for the first day at his new school. When his turn arrived to state his name,

> *I got up from my desk and hurried to the blackboard. Before the teacher could say a word, I picked up a piece of chalk and said as I wrote:*
>
> *My name is*
> *Piscine Molitor Patel,*
> *Known to all as*
> *—I double underlined the first two letters of my given name—*
> *For good measure I added*
>
> $$\pi = 3.14$$
>
> *and I drew a large circle, which I then sliced in two with a diameter, to evoke that basic lesson of geometry.*

This does the trick, and from then on, he is Pi. As he says, "In that Greek letter that looks like a shack with a corrugated tin roof, in that elusive, irrational number with which scientists try to understand the universe, I found refuge."[1]

The random-seeming sequence of digits in π echoes the curious and unpredictable currents of the sea, as Pi and Richard Parker later float on the endless blue ocean. But the digits of π are not random. There are ways to calculate them that allow us to delve as far as we like into this particular sea of numbers, many billions of digits if we wish. The true mystery of π, to me, is how it manifests itself in the most unexpected places in mathematics. We all know that π relates to circles, right? Definitely nothing to do with squares. But there's a sequence involving the square numbers 1,

4, 9, 16, 25, and so on that is connected to π in the most curious way. If we try to work out the sum $\frac{1}{1}+\frac{1}{4}+\frac{1}{9}+\frac{1}{16}+\frac{1}{25}+\cdots$, where those dots mean "carry on forever," it seems to get closer and closer to a particular value, around 1.64. This was first noticed in about 1650, and mathematicians spent more than eighty years trying to find out exactly what this number is. The great Leonhard Euler, whom we met in Chapter 2, managed to show in 1734 that, amazingly, $\frac{1}{1}+\frac{1}{4}+\frac{1}{9}+\frac{1}{16}+\frac{1}{25}+\cdots=\frac{\pi^2}{6}$. Even though I've seen the proof, it's still mind-boggling. The number π crops up in many other places, too. The equation for the famous "bell curve" in statistics involves π, and it even occurs in the meandering patterns of rivers. It's been found that if you divide the length of a river, including all its wiggles, by the as-the-crow-flies distance from source to mouth, the answer approximates to π.

Many mathematicians, from Archimedes to Newton to Charles Dodgson (better known as Lewis Carroll), have come up with ways of calculating approximations to π, but because its digits go on forever, we can never know it exactly. For Pi Patel, this is a cause of frustration. He wants things to have clearly defined endings. "What a terrible thing it is," he says, "to botch a farewell. . . . Where we can, we must give things a meaningful shape. For example—I wonder—could you tell my jumbled story in exactly one hundred chapters, not one more, not one less? . . . It's important in life to conclude things properly. Only then can you let go." Life, of course, is not tidy. All of our stories have interconnecting threads. There are no neat endings, just convenient stopping points. For Pi, after his months on the ocean with Richard Parker, perhaps it may be some consolation to know that *Life of Pi* is, in fact, precisely one hundred chapters long.

It's not just the book that has a pleasing length; Pi's time at sea is exactly 227 days. This doesn't seem, at first glance, like a particularly significant number, but I believe it is. First, if it is not, then why does Martel give us the exact number? And second, he said in an interview, when asked why he had chosen Pi's companion to be a tiger, that he had considered a rhino first, "but rhinos are herbivores and [I] didn't see how I could keep a herbivore alive for 227 days in the Pacific. So finally I settled on what now seems the natural choice, a tiger." This strongly implies that he had the 227 in mind very early on. I was so excited when I spotted why. It cannot be a coincidence that the fraction $\frac{22}{7}$ is a very good approximation to π. Unlike π, it is a rational number—we can write it as a simple, straightforward fraction. We can know it precisely. We can give it what Pi yearns for, a meaningful shape. Martel has said that he chose the name Pi because it's an irrational number, yet "scientists use this irrational number to come to a 'rational' understanding of the universe. To me, religion is a bit like that, 'irrational,' yet with it we come [to] a sound understanding of the universe." Martel's clever sleight of hand in bringing $\frac{22}{7}$ to mind with a 227-day voyage appears to achieve the impossible: it makes Pi rational.

The number π has many properties that appear paradoxical. It is irrational, yet its very definition involves a ratio: the diameter of a circle to its circumference. It is a finite number, yet its never-ending digits continue to infinity. Paradox and the infinite (and indeed paradoxes of the infinite) are a recurring theme in the work of the Argentinian author Jorge Luis Borges. His story "The Library of Babel" features a mathematical oxymoron—a finite number of things that somehow have to fill a space that extends forever in all directions. The story is a first-person account of an inhabitant of the Library—which is the universe. This "librarian" spends his life wandering the hexagonal rooms of the Library, all of which are identically laid out, reading the books and trying to understand the meaning of the cosmos. (I love the work of Borges, which is playful and profound in equal measure, and beautifully written. Please pick up a book of his short stories without delay if you've never read him.)

This particular story has an added resonance because Borges himself was a librarian—director, in fact, of the National Public Library of

Argentina. He was surrounded by books from childhood; his father had a large collection of both Spanish- and English-language works. "If I were asked to name the chief event in my life," Borges once remarked, "I should say my father's library." For such a bibliophile, it must have been a devastating loss when his eyesight started to deteriorate in his thirties; by his late fifties he had become completely blind. That knowledge makes the following sentence from "The Library of Babel" (published in 1941, when Borges was in his early forties) especially poignant: "Like all men of the Library, I have traveled in my youth; I have wandered in search of a book, perhaps the catalogue of catalogues; now that my eyes can hardly decipher what I write, I am preparing to die just a few leagues from the hexagon in which I was born."[2]

The Library is an astonishing thing, says the librarian: it contains all possible books. Every book that has been written, that is being written now, that will one day be written, that will never be written, that has been started and abandoned, that has been banned, that has been lauded, that has never been imagined exists in the Library. All the volumes in the Library are the same size, shape, and length (410 pages exactly). Already this sounds strange—but it's all right because *War and Peace*, say, can be spread over several volumes, while *The Great Gatsby* can take up part of one volume, with the remaining pages left blank.

The narrator and other librarians spend their lives wandering the Library seeking knowledge. Since all books are contained in the Library, it is inevitable that somewhere on its shelves is a book explaining how the Library came to exist, and what its structure is. There is a book telling you everything that will happen to you for the rest of your life. There is a book listing the winning numbers of every lottery on earth. There is even a copy of *Once Upon a Prime*, but given that every possible book is in the library, there are also millions of almost-copies. If the book you are holding in your hand contains any awful spelling errors or mathematical faux pas, clearly you have accidentally picked up one of these near-miss versions. As Borges's narrator says, "It suffices that a book be possible for it to exist. Only the impossible is excluded. For example: no book can be a ladder, although no doubt there are books which discuss and negate and demonstrate this possibility and others whose structure corresponds to that of a ladder."

What kind of a building could contain such a vast number of books? Here is how Borges begins his story:

> *The Universe (which others call the Library) is composed of an indefinite and perhaps infinite number of hexagonal galleries, with vast air shafts between, surrounded by very low railings. From any of the hexagons one can see, interminably, the upper and lower floors. The distribution of the galleries is invariable. Twenty shelves, five long shelves per side, cover all the sides except two. . . . One of the free sides leads to a narrow hallway which opens on to another gallery, identical to the first and to all the rest. To the left and the right of the hallway there are two very small closets. [These are for sleeping and other physical requirements.] Also through here passes a spiral stairway, which sinks abysmally and soars upwards to remote distances.*

Every room of the Library, then, has the same design, and the Library continues indefinitely in all directions. This presents a problem, because even though the number of possible books is almost unimaginably vast, it is, as I hope to convince you shortly, nevertheless finite. (There's a lot going on, mathematically speaking, in Borges's story, so much so that the mathematician William Bloch wrote an entire book about it. But I want to focus just on this main paradox.)

Can the librarians be correct that the entire Library is of the unending structure described and that at the same time every possible book appears in the Library just once, with no duplicate copies? Let's explore this idea a little more. Borges gives us a bit more to go on about the contents of each room and the shape and size of each book. Each hexagonal room, says the librarian, has bookshelves on four of its walls—we must leave room for the entry and exit, after all. These four walls each contain five bookshelves. Each shelf holds 32 books. (Just a note here for those reading Borges in English translation: at least one translation has 35, not 32, but I did check the Spanish version and it says 32, so I'm sticking with that.) A bit of mental arithmetic later, and we find that each room of the Library contains exactly $4 \times 5 \times 32 = 640$ books.

Now for the harder question: How many books are there in the Library? We need a little more information from the story. All the books, says the librarian, are identical in their format. Each has 410 pages. Each page has 40 lines, each line has 80 characters, and there are 25 characters, comprising 22 alphabet letters, along with the comma, the period, and the space. Borges doesn't tell us precisely what the alphabet letters are. Obviously, this isn't the English alphabet with its 26 letters, nor the Spanish one, which has all ours plus *ñ*. Depending on whom you ask, the classical Latin alphabet had between 21 and 23 letters, so perhaps this is what Borges had in mind. In any case, with 80 characters in the 40 lines of 410 pages, that's a total of $80 \times 40 \times 410 = 1{,}312{,}000$ characters. I don't know about you, but I definitely needed a calculator for that one. In the story, the librarian says that there are characters on the spines of the books as well. We don't know how many, but since the writing on book spines is usually written vertically, a reasonable assumption, given that the pages inside have 40 lines, is that there is room for 40 characters on the spine. There are twenty-five choices for each character. This is a bit like our calculations with limericks and sonnets earlier, but just as a reminder: imagine that our books just use 3 characters, *a*, *b*, and *c*, and imagine that each book is just 2 characters long. We have three choices for the first character, *a*, *b*, or *c*. When adding the second character, each of these three can be followed by three choices for the second character, which means there are $3 \times 3 = 9$ possibilities. Here they are:

aa *ba* *ca* *ab* *bb* *cb* *ac* *bc* *cc*

If we add another letter, then each of the 3^2 possibilities for the first two letters has three ways to add a third letter. So the total is $3^3 = 27$. Here they are:

aaa	*baa*	*caa*	*aba*	*bba*	*cba*	*aca*	*bca*	*cca*
aab	*bab*	*cab*	*abb*	*bbb*	*cbb*	*acb*	*bcb*	*ccb*
aac	*bac*	*cac*	*abc*	*bbc*	*cbc*	*acc*	*bcc*	*ccc*

If each book with a three-letter alphabet had seven letters in total, then there would be $3^7 = 3 \times 3 \times 3 \times 3 \times 3 \times 3 \times 3$ possible books. I thought I was picking those numbers randomly, but I realize that in 3 and 7 I have chosen two of the most ingrained pattern numbers in Western thought—I guess there's no escaping. If a three-letter alphabet book has n letters, then there are 3^n such books. By exactly the same argument, the number of Babel-style books with n characters and twenty-five possibilities for each one (including spaces, commas, and periods, remember) is 25^n. Since each book contains 1,312,000 characters, there are a whopping $25^{1,312,000}$ different possibilities for the contents of a book. We mustn't forget about the spine, either. We are assuming that there are forty additional characters on the spine. This means there are $25^{1,312,040}$ books in the Library of Babel.

At this point, a calculator isn't going to cut it. In fact, even a computer is going to be fairly useless, because $25^{1,312,040}$ is a ridiculously vast number. The nearest power of ten to it is $10^{1,839,153}$, which is 1 followed by 1,839,153 zeros. Writing down all those zeros, if you could write a respectable five every second, would take 102 hours. More importantly, this number proves categorically that the Library of Babel cannot be part of our universe. In our universe, scientists estimate that there are "only" 10^{80} atoms, so unless you can think of a way to fit untold billions of books onto every atom, the Library universe must be different from our own, and much bigger.

Whatever universe we postulate, there is a problem. We have $25^{1,312,040}$ books, and they are contained in a series of identical rooms each of which contains exactly 640 books. To find out how many rooms there are in the Library, then, we just have to divide $25^{1,312,040}$ by 640. But $25^{1,312,040}$ is a bunch of 25s multiplied together, and 25 is an odd number. If you multiply a load of odd numbers together, the outcome at the end is still an odd number. In this case, it's an unimaginably vast one, but it's still odd. And you can't divide an odd number by 2 and end up with a whole number; that's more or less the definition of an odd number. We don't have to actually do the calculation, then, to know for sure that $25^{1,312,040} \div 640$ is not a whole number. But this would mean that the Library doesn't have a whole number of rooms!

In his book, Bloch suggests that one way to fix this might be to tweak the numbers from the story. He shows that a whole number of rooms will result if we let each shelf have forty-nine books rather than thirty-two, and change the number of characters allowed to twenty-eight. But I prefer as far as humanly possible to follow the rules given in the story, otherwise what's the point? We do have a tiny bit of ambiguity to cling to, and that's the writing on the spines of the books. We have decided that there are forty characters on the spine of each book (and this can include spaces). Changing the number of characters actually won't solve the problem, though, because however many there are, we still end up with lots of 25s multiplied together, which is still an odd number. However, I have two suggestions that I hope respect the Babel universe. The first is that because book titles don't usually have periods, we could assume that not twenty-five but twenty-four characters are allowed for book spines. This would mean that there are 24^{40} possible book spines, and we already know that there are $25^{1,312,000}$ possible book contents. So the total number of books in the Library would be $25^{1,312,000} \times 24^{40}$. This, at least, is an even number. And it is in fact exactly divisible by 640. Remember that an expression like 24^{40} means a string of forty 24s multiplied together. We can break that apart however we like—for instance, a string of seven 24s followed by thirty-three 24s, all multiplied together. In other words, $24^{40} = 24^7 \times 24^{33}$. Bear with me while I do a little calculation:

$$25^{1,312,000} \times 24^{40} = 25 \times 25^{1,311,999} \times 24^{7} \times 24^{33}$$
$$= (25 \times 24^{7}) \times (25^{1,311,999} \times 24^{33})$$
$$= 114,661,785,600 \times (25^{1,311,999} \times 24^{33})$$
$$= 179,159,040 \times 640 \times (25^{1,311,999} \times 24^{33})$$

Aha! This huge number is a multiple of 640. This means that the Library fits exactly into a whole number of those hexagonal rooms, namely $179,159,040 \times 25^{1,311,999} \times 24^{33}$.

My other suggestion is based on something in the story itself. When the librarian is explaining the rules of the Library, he tells us that "the best volume of the many hexagons under my administration is entitled *The Combed Thunderclap* and another *The Plaster Cramp* and another *Axaxaxas mlö*." By the way, this last one is an in-joke that references one of Borges's other stories, "Tlön, Uqbar, Orbis Tertius." The phrase *axaxaxas mlö* means something like "the moon rose" in the language of a planet—Tlön—that may or may not be real, the only evidence for whose existence is snatches of information found in certain copies of certain books. Since all possible books exist in the Library of Babel, it must certainly contain all works by Borges, and all literature from Tlön, whether Tlön exists or not. In any case, since one of the book spines contains the accented letter *ö*, then perhaps the choice for spine letters is greater than twenty-five. If we allow just one extra letter, so that there are twenty-six choices for each spine letter, then we get a factor of 26^{40} in our calculation of the number of books, and the upshot, again, is a total number that is exactly divisible by 640. Either way, I think we can stick closely enough to the spirit of the story and still end up with a whole number of hexagonal rooms.

Having established that there is a very large, but still finite, number of rooms in the Library, the big question is: How do we square this with the fact that the Library is supposed to continue indefinitely in all directions? Can mathematics help us figure out a possible structure that has all the qualities stated in the text? For instance, we have these ventilation shafts that pass through the center of each hexagon, upward and downward forever. There are also staircases spiraling up and down from corridors between hexagons. What that tells us is that the configuration of hexagons and staircases must be exactly the same on each vertical level.

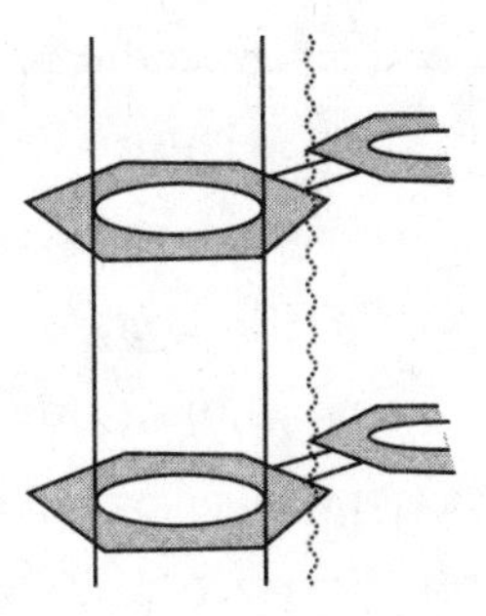

We also know that exactly two of the walls of the hexagon do not contain bookshelves. One or both of these walls lead to a corridor linking two hexagons together horizontally (it also connects to a spiral staircase). All hexagons are identical, so either just one wall of every hexagon leads to a corridor, or two walls do. The first possibility is not going to work. The reason is that it would isolate pairs of hexagons on any level. If Hexagon A links to Hexagon B, then Hexagon B already has the corridor from Hexagon A, and so can't be joined to any other hexagon. But the story tells of traveling "miles to the right" and "90 levels up," for instance, so we cannot have just two hexagons on each level. So we had better have two corridors leading from each hexagon. One possibility is what I've shown here, with the corridors leading off opposite walls. Then each horizontal layer is a chain of hexagons in a line:

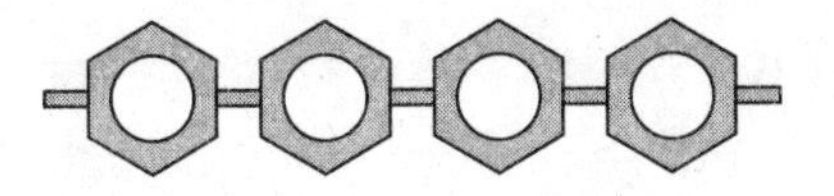

But perhaps the corridors are on adjacent walls, or have one bookcase wall between them rather than two. If that's the case, then there are many more possibilities for the floor plan on each level, which I invite you to explore. Let's assume for the moment, though, that each horizontal level is a line of linked hexagons, replicated above and below—I picture a kind of huge rectangular chain mail mesh of hexagons.

This is all very well, but we are supposed to be able to move around forever, up, down, right, and left, without reaching the end. There is a shape that, although finite, has no end and no beginning. We use it to symbolize eternal love when we wear wedding rings: the circle. Moving around a circle we can carry on forever, and any point feels like any other. This is a one-dimensional "infinite line" that fits in a finite space. Going up a dimension, we can walk all over the surface of a sphere (such as our own planet) without ever reaching its end or falling off the edge. A sphere big enough to hold all the books in the Library of Babel would certainly not be something you could circumnavigate in a lifetime, so it would feel infinite while still being finite.

But our rectangular mesh of rooms can't quite be sitting on the surface of a three-dimensional sphere—if you try to wrap a rectangular piece of paper around a ball you'll see the problem. Some parts inevitably get (technical term) smooshed together. This is why mapmaking is such a challenging art—you can't draw a map of the spherical earth without distortion. One solution to the challenge is to go up a dimension. The Library universe could exist as the three-dimensional surface of a four-dimensional sphere! Mathematically speaking, this works nicely, but there's another possibility that I like better. It may indicate a misspent youth, but my favorite contender for the shape of the Library of Babel is what you might call the "space invaders" solution. In days of yore when early computers had the memory capacity of an amnesiac goldfish, there were lots of alien-zapping games that had you moving through space shooting enemy craft, and, presumably to save memory, if your ship exited from the right-hand edge of the screen, it would reappear at the corresponding point on the left-hand edge, as if these were the same point in space. Or you might find that flying off the bottom of the screen caused you to reappear at the top of the screen.

Mathematicians do this sort of thing all the time in the area of mathematics known as *topology*. You just decree that, for instance, the bottom boundary of the screen is exactly the same set of points as the top boundary and identify those two edges with each other. At the cost of a bit of distortion, we can make this happen for real in three dimensions—turning a flat surface into a slightly curved one by just curving the rectangle around and gluing the two edges together to make a cylinder.

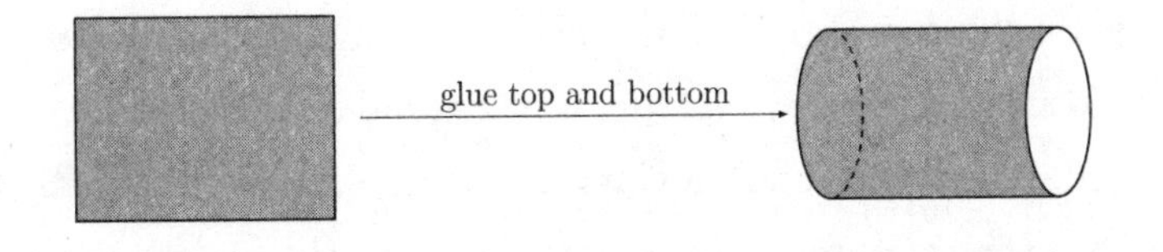

Trying this with our mesh of hexagonal rooms works nicely. The vertical layers now form a vast circle, and we could move up and down between layers and never come to an endpoint. Nor would we find ourselves turning upside down; after all, people on the other side of the world from you are not all standing on their heads. But what about the horizontal layers? They still have a boundary, a "last hexagon," at the far left and far right of the cylinder. Mathematicians never like to let a good idea go to waste. We just use exactly the same trick again for the horizontal layers. That is, when our spaceship leaves the screen on the right, it enters again on the left. We mathematically "glue together" the two ends of the cylinder. We can visualize this as creating what's called a *torus* in the world of math and a doughnut everywhere else. (Though not a British doughnut, as these are filled with jam and would get the books sticky.)

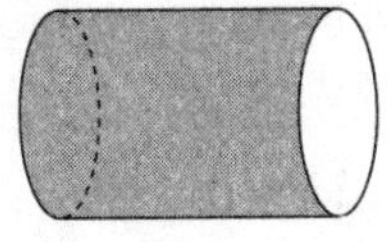

glue left and right

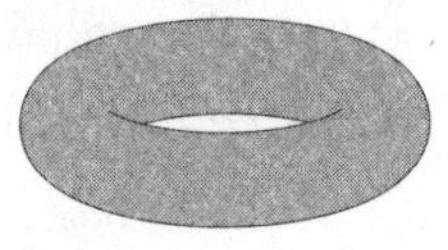

The final author I want to talk about in this chapter is someone who would definitely have enjoyed the idea of a doughnut-shaped universe filled with all possible books. Let us leave our wandering librarians, then, and take a trip to Wonderland.

The most famous works of fiction written by a mathematician are surely Lewis Carroll's *Alice's Adventures in Wonderland* and its sequel, *Through the Looking-Glass.* Their playful mathematics and mind-bending logic enhance the surreal and dreamlike quality of Alice's imagined worlds. For me, although there are a good many overtly mathematical references in his work, it's his entire approach to storytelling that reveals his mathematical cast of mind. Lewis Carroll, as you may know, was the pen name of the Reverend Charles Lutwidge Dodgson, a mathematician and clergyman who lived and worked at Christ Church College, Oxford, in the latter half of the nineteenth century. The Latinized version of Charles Lutwidge is Carolus Ludovicus, from which it's a short step to Lewis Carroll.

All his fiction and poetry has a reductio ad absurdum flavor to it that in fact is common to both mathematics and children's games of make-believe. If, for example, we assume that you can grow and shrink at will (or by eating cake and drinking potion), then it would be possible to swim in a lake of your own tears, as Alice does shortly after falling down the rabbit hole. You observe the internal logic of the game—precisely as mathematicians do. We agree on the ground rules of our mathematical playground, and then we explore.

Mathematically, pushing assumptions to their logical limits in the hope that they'll break is a common proof technique. The trick is to assume the opposite of what you think is true. Already it's a bit *Through the Looking-Glass.* This was how we proved way back in Chapter 1 that there are infinitely many prime numbers. We assumed there weren't, in which case there would be a finite list containing all the prime numbers, and then we deduced from that the existence of a prime number not on the list, which would be impossible. This is a true mathematical reductio ad absurdum, a trick we call "proof by contradiction." In a similar vein, Alice's encounter with the Mock Turtle features, as well as some very silly puns, a sequence that Alice tries to take to its logical conclusion. The Mock Turtle is recounting his school days, when they studied "all the different branches of Arithmetic—Ambition, Distraction, Uglification, and Derision."

"And how many hours a day did you do lessons?" said Alice. . . .

"Ten hours the first day," said the Mock Turtle: "nine the next, and so on."

"What a curious plan!" exclaimed Alice.

"That's the reason they're called lessons," the Gryphon remarked: "because they lessen from day to day."

This was quite a new idea to Alice, and she thought it over a little before she made her next remark. "Then the eleventh day must have been a holiday?"

"Of course it was," said the Mock Turtle.

"And how did you manage on the twelfth?" Alice went on eagerly.

"That's enough about lessons," the Gryphon interrupted in a very decided tone.

And no wonder, because in this plan, they would have had to learn for fewer than zero hours on every subsequent day.

There is a good deal of absurd arithmetic in Lewis Carroll's verse too. In his poem "The Hunting of the Snark" (subtitled "An Agony in 8 Fits"), ten crew members, all of whose names begin with *b*, set out on a voyage to track down the snark, though ultimately they fail because it turns out that all along the snark was a boojum. At one point the Beaver is struggling to work out how to add two to one to get three. The Butcher steps in to help. He "explained all the while in a popular style, which the Beaver could well understand":

Taking Three as the subject to reason about—
A convenient number to state—
We add Seven, and Ten, and then multiply out
By One Thousand diminished by Eight.
The result we proceed to divide, as you see
By Nine Hundred and Ninety and Two:
Then subtract Seventeen, and the answer must be
Exactly and perfectly true.

At first glance this looks, to use a technical term, like a load of nonsense. But in fact it is a clever little mathematical trick, a series of precise logical steps that leads inexorably (if ridiculously) to the right answer. The Butcher is trying to show that 3 is the answer to the fiendish sum $2+1$. So he starts with 3 and then does a bunch of arithmetical calculations that if you follow them through carefully, actually bring you back precisely to 3 again. But even better, this works whatever number you start with. If I start with my favorite number, 4, I'll end up with 4. Try it: We add 7 and 10 to our number. So we have $4+17=21$. Then multiply this by $1000-8$, which is 992. Then we divide by 992. So far, then, we have found $(4+17)\times\frac{992}{992}$, which is just $4+17$. The final instruction is to subtract 17, which brings us back to 4. Wherever you start, the answer must indeed be exactly and perfectly true.

There's one specific number that Lewis Carroll seems to have been a little obsessed with: 42. It crops up all over the place in his writing. In *Alice's Adventures in Wonderland* (which happens to have forty-two illustrations), the King of Hearts, infuriated by the rapidly enlarging Alice's disruption of court proceedings, reads out from his notebook, "Rule Forty-two. All persons more than a mile high to leave the court." When Alice follows the White Rabbit into his hole, she tumbles down a very deep well, and keeps falling down and down. She wonders if she will fall right through the earth. It's a curious mathematical fact that falling through a tunnel leading between any two points on the earth's surface takes a constant amount of time (being a pure mathematician, I ignore prosaic things like friction and air resistance). Guess how long it would take Alice to fall all the way through the earth to the other side? You got it: forty-two minutes.

There are other possible 42s hidden in *Through the Looking-Glass*. While *Alice's Adventures in Wonderland* is full of playing cards, in *Through the Looking-Glass* the theme is chess. The entire book has the structure of a chess game, white versus red, with Alice moving through a board laid out over the fields. Alice meets several of the pieces during the course of her adventure, which it's possible, says Lewis Carroll, to play as a real game of chess with Alice as a pawn who crosses the board to become a queen. During one conversation, Alice tells the White Queen she is 7 and a half years old exactly, or 7 years, 6 months; 7 times 6 is, of course, 42. The Queen's age is much greater: 101 years, 5 months, and a day. How many days is that? The answer depends on where your leap years go, but the highest attainable total is 37,044. Is this a randomly chosen number? Perhaps. But the Red Queen and the White Queen, being from the same chess set, are presumably the same age. So their combined age is 74,088 days. What of that? Well, it's exactly 42 × 42 × 42. I struggle to believe that this is a coincidence.

I've never read a convincing explanation as to why Lewis Carroll was so fixated on 42. In spite of his enthusiasm for logic, I suspect he just took a shine to it. But there is one school of thought that says there might be a religious interpretation. In the preface to "The Hunting of the Snark," for example, we hear how the crew's rules had entangled them in a logical impasse:

> *Rule 42 of the Code, "No one shall speak to the Man at the Helm," had been completed by the Bellman himself with the words "and the Man at the Helm shall speak to no one." So remonstrance was impossible [and] during these bewildering intervals the ship usually sailed backwards.*

Some people believe that the number 42 is a reference to an important religious document, the *Forty-Two Articles* of Thomas Cranmer, which laid out important doctrinal rules of the Church of England. Lewis Carroll was an Anglican priest, so he would certainly have been familiar with this document. Article 42, by the way, is "All men shall not be saved at the length." Make of that what you will.

The number 42 has become much better known in the last forty-two years (or thereabouts: the TV series appeared in 1981) for its role in Douglas Adams's book *The Hitchhiker's Guide to the Galaxy*. He may have been inspired by Lewis Carroll—after all, the episodes of the original radio series, on which the TV series and books were based, were named Fit the First, Fit the Second, and so on, just like the parts of "The Hunting of the Snark." In *The Hitchhiker's Guide*, an alien civilization of hyperbeings creates a huge computer called Deep Thought that takes seven and a half million years (a million Alice ages) to determine the answer to "Life, the Universe, and Everything." After those eons have passed, Deep Thought reveals that the ultimate answer is 42. The problem then becomes to work out, in a sort of existential version of *Jeopardy!*, what is the question?

Let's address a final arithmetical mystery. This one combines numbers with Carroll's favorite pastime: setting up a mathematical chain of events that leads down a logical rabbit hole. When Alice arrives in Wonderland, she starts to doubt her own sanity because everything is so confusing. She decides to see if she still knows reliable things like multiplication tables. "Let me see: four times five is twelve, and four times six is thirteen, and four times seven is—oh dear! I shall never get to twenty at that rate!" Poor Alice—but what does she mean, she'll never get to twenty? The prosaic interpretation is that traditionally we learn our times tables only up to twelve, and, following her pattern, if $4 \times 5 = 12$ and $4 \times 6 = 13$, then $4 \times 7 = 14$, $4 \times 8 = 15$, $4 \times 9 = 16$, $4 \times 10 = 17$, $4 \times 11 = 18$, and $4 \times 12 = 19$. Since we stop at twelve, we won't reach $4 \times 13 = 20$.

But there's a much more mathematically interesting interpretation, and that's trying to find a scenario in which 4 times 5 really is 12. This isn't as ridiculous as it seems when you remember that clocks follow an arithmetic in which 6 plus 8 is 2. What I mean is, if you add eight hours to six o'clock, you don't arrive at fourteen o'clock (unless you are in the military and have to use the twenty-four-hour clock), but two o'clock. In some situations, then, it's legitimate to say that $6+8=2$.

Another way to get surprising answers to sums is to work in an unexpected base. In our usual base 10, we write numbers in terms of powers of 10 (units, 10s, 100s, 1,000s, and so on), so that 1101 in base 10 means one thousand one hundred and one. But in binary or "base 2" arithmetic, we work in powers of 2 (units, 2s, 4s, 8s, and so on). This time, 1101 means 8 plus 4 plus 1: in other words, 1101 equals 13. And we can write sums that look crazy but are correct: $1 + 1 = 10$, or $10 + 1 + 1 = 100$. Computer programmers occasionally use a base 16 system (hexadecimal). In hexadecimal, 14, say, would mean 1 lot of 16 plus 4 units. In other words, 20. So in hexadecimal, we can correctly write that $4 \times 5 = 14$. The game now becomes, in what number base is it true that $4 \times 5 = 12$? It turns out the answer is base 18, because in base 18, 12 means one lot of 18 plus 2, which indeed is twenty. What about $4 \times 6 = 13$? Here we need base 21, because 4×6 is 24, and 13 in base 21 means one lot of 21 plus 3, which is the required 24. This pattern continues nicely if we add 3 each time to the base. We get

$$4 \times 7 = 14 \text{ (base 24)}$$
$$4 \times 8 = 15 \text{ (base 27)}$$
$$4 \times 9 = 16 \text{ (base 30)}$$

The pattern continues up to 4 times 12, which indeed does equal 19 in base 39 (one lot of 39 plus 9). But frabjous day! Callooh! Callay! We can never get to 20 like this. Four times 13 is 52, and to fit the pattern the next base should be 42 (there it is again). But in base 42, writing "20" would represent two 42's, which is 84. So we really will never get to 20. I especially like that the pattern breaks not only at base 42, but also when the total reaches 52—which is the number of playing cards in a pack. It's a nice reference to the later appearance of cards, like the Queen of Hearts, as characters in the story.

The examples I've shown you are glimpses of the common thread that runs through all Lewis Carroll's writing, both mathematical and nonmathematical: a drive to understand the power and possibilities of logic. As well as his children's books, he invented games and puzzles, many for children, that were aimed at teaching the laws of logical inference, starting with the most basic syllogisms (All men are mortal; Socrates is a man; therefore Socrates is mortal) and moving to sequences of deductions with as many as a dozen or more sentences strung together.

Carroll's fiction, including the *Alice* books, is for me simply another facet of his lifelong exploration into how far we can get by setting the scene and then following the logic. The discussions of words and their meanings in the *Alice* books is one telltale sign of these mathematical undercurrents. For a mathematician, there is resonance in Humpty Dumpty's remark "When I use a word it means just what I choose it to mean—neither more nor less." In mathematics we have to be absolutely clear about the meanings of the words we are using, and must not load them with unspoken qualities. It's not mere pedantry—any ambiguities risk tying us in logical knots and can even mean our deductions are false. It doesn't matter what names we give to our new concepts, but we have to be careful to get the definitions right. As I mentioned earlier, all sorts of things go wrong if our definition of prime numbers allows 1 to be prime. Just like Humpty Dumpty, a mathematician's words must mean no more, or less, than what they say.

Lewis Carroll, along with other Victorian mathematicians like John Venn (of Venn diagram fame), was interested in taking this precision further and codifying the very processes of logic itself. This "symbolic logic" allows you not just to look at whether individual statements are true or false, but to make deductions about the truth or falsity of statements made by connecting them with words like "and," "or," or "implies." Even these simple words can trip us up if we aren't careful. The word "or," for instance, can mean different things according to context. Don't believe me? Would you like a cup of tea or a cup of coffee? In that sentence, we all know "both" isn't encompassed in "or." On the other hand, a job advertisement saying that applicants should be fluent in Spanish or Portuguese would presumably not exclude people fluent in both languages. In normal speech we can tell from context which meaning is intended. If you are trying to create a set of rules of logic that covers all possibilities, you don't have that luxury.

The "symbolic" bit of "symbolic logic" comes from the fact that we use symbols for words like "or" and "and," and with them we construct a sort of algebra of logic. The aim is to be able to extract all possible logical conclusions from a collection of statements. One example given by Carroll is these two sentences: "No son of mine is dishonest" and "All honest men are treated with respect." Now, we don't concern ourselves with determining whether these sentences are true or false. Our job is to say what can be deduced, assuming they are true. Carroll describes how this is just one instance of a more general archetype, of the form "No x is not y, and every y is z." If both these are true, then it must follow that "no x is not z." Carroll represents these ideas both with diagrams and with symbols. In symbolic form, using his notation, we have the rather forbidding $xy'_0 \dagger yz'_0$ ¶ xz'_0 (here, † means "and," and ¶ means "therefore"). Once we know this general formula, we can apply it in our special case, where x is "my sons," y is "honest," and z is "treated with respect." And hey presto, we can deduce that "no son of mine fails to be treated with respect." Carroll assures us that it gets easier with practice!

This example comes from a book that Lewis Carroll wrote, aimed at popularizing symbolic logic for the general public. In the introduction, he extols its virtues thus:

> *Mental recreation is a thing that we all of us need for our mental health. . . . Once master the machinery of Symbolic Logic, and you have a mental occupation always at hand, of absorbing interest. . . . It*

> *will give you . . . the power to detect fallacies, and to tear to pieces the flimsy illogical arguments, which you will so continually encounter in books, in newspapers, in speeches, and even in sermons, and which so easily delude those who have never taken the trouble to master this fascinating Art. Try it. That is all I ask of you!*

Lewis Carroll's contributions to the academic study of symbolic logic were valuable and important. As befits his character, he was adorably enthusiastic too about bringing the joys of that subject to the masses. In spite of his valiant efforts, though, I'm sorry to say that it did not really catch on as a fun family pastime.

I can't resist ending this chapter with a story that may be apocryphal but is so good it deserves to be true. Believing unlikely things is, after all, just a matter of practice, says the White Queen: "When I was your age, I always did it for half-an-hour a day. Why, sometimes I've believed as many as six impossible things before breakfast." Legend has it, anyway, that Queen Victoria was so delighted with *Alice* that she asked to be sent the very next book Mr. Carroll produced. History does not relate her reaction on receipt of *An Elementary Treatise on Determinants, with Their Application to Simultaneous Linear Equations and Algebraical Geometry*. One suspects she was not amused.

10

Moriarty Was a Mathematician

The Role of the Mathematical Genius in Literature

In the bestselling *Millennium* series, there's a scene at the start of the second book in which the hero, Lisbeth Salander, comes up with a short proof of Fermat's Last Theorem. This may be the most famous misnomer in mathematical history. The mathematician Pierre de Fermat, undoubted genius though he was, made a great many claims that he didn't back up with proof. This "theorem" of his was one such statement. In mathematics, we call these things *conjectures*. Most of Fermat's conjectures were resolved either by Fermat himself or by other mathematicians within a few years. But this particular one was not—hence the "last." What made it even more enticing was the accompanying marginal note—"I have a truly marvelous proof, but this margin is not big enough to contain it." Hundreds of mathematicians tried to find this proof, but as the decades stretched into centuries, nobody succeeded. Even partial progress involved major new mathematical advances, far beyond anything Fermat could have had in mind. Eventually, in one of the great achievements of the last half century, the problem was cracked, and a proof found, by Andrew Wiles in 1993, using brilliant, beautiful, incredibly sophisticated mathematical machinery.

Anyway—we are asked to believe that Lisbeth Salander, genius hacker with no mathematical training, has proved what I propose should be called not "Fermat's Last Theorem" but "Fermat's midlife boast." This gambit, of course, is shorthand for Salander's being a maverick genius, with perhaps a side order of being good only at logic, not human emotion. The author might just as well have written, "Insert evidence of extreme cleverness here."[1]

In this final chapter of the book, I'm going to show you some of the ways that people who do mathematics have been portrayed in literature. All too often, as mentioned briefly in Chapter 8, we see the trope of the emotionless, uncaring, obsessive, even insane mathematician. This stereotyped portrayal does mathematics a disservice, perpetuating the idea that only "freak" geniuses can be mathematicians when really everybody can delight in the fascination of mathematics. More sympathetic portrayals are out there, and I'll show you some of these too, from Aldous Huxley's heartrending "Young Archimedes" to Alice Munro's *Too Much Happiness*, a mesmerizing fictionalized account of the life and death of the Russian mathematician Sofya Kovalevskaya.

Let's begin our account with the most straightforward, if unrealistic, type of mathematician in literature: the character motivated entirely by logic, untrammeled by messy things like emotion. In Isaac Asimov's much-loved *Foundation* novels, a mathematician named Hari Seldon uses a new field of probability theory called *psychohistory* to predict the future of the galaxy. The seemingly stable galactic empire will fall, and there will be thirty thousand years of chaos. But if we use mathematics, we can reduce that ages-long darkness to just one millennium.

What I think appeals to a lot of people in Asimov's books is the beguiling fantasy that scientists, and especially mathematicians, are driven by pure reason, that cleverness can get you out of any fix, and that everything can finally make sense if you can just ramify the ninth-dimensional asymptotes over a tangential vector field. Sadly, you can't, first, because life isn't like that, and second, because I've just made up all those phrases, so they are meaningless. These books don't have great dialogue, and the characters are not three-dimensional, but that's not their point. It's all about the ideas. Hari Seldon exists to explain the ideas. He doesn't need a backstory. The big-budget TV adaptation released in 2021 tried a bit harder with backstory, not always successfully. I did find myself scoffing at the young math genius who lists primes when she's nervous, until I remembered that when, as a teenager, I had to walk past the bus stop where boys from the local school congregated to catcall passing girls, I used to construct the rows of Pascal's triangle in my head to stay calm.

A mathematician like Hari Seldon, in fiction, is less of a character and more of a plot device, a being of perfect logic. You get the feeling that if they do the right thing, it's only because it happens to coincide with doing the logical thing. They are essentially amoral. If the equation had another solution, then they could just as easily turn into the villain of the piece.

On that note, let's meet Professor James Moriarty, the "Napoleon of crime," nemesis of Sherlock Holmes. He is a man "endowed by nature with a phenomenal mathematical faculty" and apparently an expert in the "Binomial Theorem." He wrote a treatise on it that enjoyed a "European vogue" and won him "the Mathematical Chair at one of our smaller universities." The binomial theorem is a real theorem, but I have to say that since it was, even by that time, a rather elementary result in pure mathematics, this is as silly as portraying someone as a professor of adverbs. An academic career is not, ultimately, for Moriarty—instead he decides to use his vast intellect to become a criminal mastermind.

But there's something about this characterization that bothers me. Holmes, after all, is something of a mathematician himself—he has written a treatise on cryptography, for example. He venerates pure logic, chiding Watson for bringing emotions into things: "Detection is, or ought to be, an exact science. . . . You have attempted to tinge it with romanticism, which produces much the same effect as if you worked a love-story or an elopement into the fifth proposition of Euclid." So why is it the evil Moriarty who is a mathematician rather than the logic-obsessed Holmes? My suspicion is that it's because of the stereotype of the mathematician as a mere calculating machine. That phrase, in fact, was used by Conan Doyle in his initial version of Holmes, "but I had to make him more of an educated human being as I went on with him." He must become human, or we won't form an emotional attachment to him.

Moriarty, by contrast, was created for a single purpose: to kill off Holmes. He makes his first appearance in what Conan Doyle, tiring of writing detective stories, had intended to be the last Holmes mystery: "The Final Problem." Moriarty doesn't need to be a human being. He is simply the perfect anti-Holmes: his true equal in intellect and therefore the only person who can kill him. Since they are mathematically equal, the only possible outcome occurs: they cancel each other out, plunging to their deaths together over the Reichenbach Falls.

Or so we thought. Naturally, the fan base revolted. Twenty thousand people canceled their subscription to *The Strand Magazine*, where the stories were published, and Conan Doyle received hundreds of anguished, pleading, even angry letters (including one from a distraught lady that began "You brute"). After eight long years, a period known to fans as the Great Hiatus, Conan Doyle eventually gave in to the pressure and came back with a real humdinger (*The Hound of the Baskervilles*). He would go on to write more than thirty additional Holmes stories, several of which also feature Moriarty.

Literature has its fair share of tortured geniuses (the chess prodigy Beth Harmon, in *The Queen's Gambit* by Walter Tevis, is just one example), so it's no surprise that mathematicians get this treatment too.[2] Aldous Huxley is best known for his dystopian novel *Brave New World*, but in 1924 he wrote a poignant story about a mathematical child prodigy. In "Young Archimedes," the narrator recounts how, during a stay in an Italian villa, his young son, Robin, befriends a local peasant boy, Guido, a thoughtful child "given to sudden abstractions." Guido loves music, and, seeing this, the narrator starts to teach him to play the piano, at which he shows great aptitude. But that's not Guido's true passion. One day, the narrator happens to spot the two boys drawing in the sand, and to his astonishment he finds that Guido has discovered Pythagoras's theorem for himself and is showing Robin a proof of it. (Robin couldn't be less interested, and makes Guido rub it out and draw a picture of a train instead.) That's part of the tragedy—that nobody around this young boy can make sense of the beauty that he sees in mathematics. The narrator begins exploring geometry with Guido, and even teaching him algebra. Guido's delight in it is wonderful. But then it is all taken away. The villa's owner, Signora Bondi, prevails upon Guido's father to allow her to take him away to train him as a pianist. She confiscates the geometry books and forbids him to do mathematics, and Guido's chance of becoming a great and more important, fulfilled, mathematician is gone. It reminds me of those famous lines from Gray's "Elegy": "Full many a flower is born to blush unseen / And waste its sweetness on the desert air."

In "Young Archimedes," Huxley remarks that child prodigies are usually either musical or mathematical, or both: "Till he was thirty Balzac gave proof of nothing but ineptitude; but at four the young Mozart was already a musician, and some of Pascal's most brilliant work was done before he was out of his teens." I'm not sure how true this is—it seems a bit harsh on poor Balzac—but for me the most fundamental thing that music and mathematics have in common (and chess too—another arena in which child prodigies appear) is pattern. All human beings have an innate appreciation of pattern, and at the extremes, an ability to pick up on and mimic patterns can get you a long way in both mathematics and music. You don't have to understand a Mozart sonata to be technically capable of playing it, and you don't have to understand an equation to learn the algorithm that solves it. You can get a fair distance by spotting patterns and learning tricks, and that might be why it's more possible to be a prodigy in these disciplines. Some of these child geniuses do go on to be exceptional mathematicians—or musicians; most do not, and that's fine. I want everyone to enjoy mathematics, just as everyone can enjoy music, independent of our level of skill. To say it's pointless doing it unless you are going to be amazing at it is as stupid as saying nobody should play sports except Olympic athletes.

To give him his due, Huxley does not portray Guido as just a learner of tricks, a reciter of the digits of π, but a true mathematician taking a deep joy in discovery. Still, the idea of mathematics as a "strange distinct talent," as Huxley puts it, is an invidious one. Human beings are mathematical beings, and we can all engage with mathematical ideas. It's not true that you either have it or you don't, nor that if you aren't already a genius when young, there's no hope for you. Unfortunately, the gatekeepers of mathematics have not always felt that way. In 1940 the English mathematician G. H. Hardy wrote *A Mathematician's Apology*, describing his views on what mathematics is and why it's important. There's a lot I like about the book—he is very eloquent in his description of mathematics as a creative art like poetry or painting.[3] But he opens by saying that even to be writing such a piece is an admission that he's past it, as a mathematician, because exposition is "for second-rate minds." Wow. And don't try to do mathematics if you're over forty, or (perish the thought) a woman, because mathematics, he says, "is a young man's game."

What happens when a child prodigy grows up is a question explored in the charming and funny novel *Uncle Petros and Goldbach's Conjecture* by Apostolos Doxiadis. Published in English in 2000 (a reworking by Doxiadis of his earlier Greek-language version), it tells the story, through the eyes of his nephew, of the eponymous Uncle Petros and his doomed attempts to prove a famous mathematical conjecture. Featuring entertaining cameo appearances by real mathematicians, including Hardy, what the book does really well is to capture something of the emotional experience of actually doing mathematical research. Apostolos Doxiadis studied mathematics at university, and it shows.[4] It's not that there is a lot of complicated algebra in the book, more that the descriptions of life as a working mathematician ring very true. Going into battle with a theorem can involve months or even years of frustration, as you try over and over again to find the key idea that makes everything work. Sometimes, eventually, inspiration strikes, and you make progress. Those days are exhilarating and make everything worthwhile.

Occasionally, at times of despondency and fatigue, your brain lies to you and makes you think you have solved the problem, just to get you to stop work for the day: "Petros now often had the feeling that he was almost a hair's breadth away from the proof. There were actually even a few exhilarating minutes, late on a sunny January afternoon, when he had the short-lived illusion that he had succeeded." This happens to every mathematician—the only thing to do is step away from your work before you see the inevitable error, and take a few hours' rest. Who knows, you might dream the proof that night. (I've experienced this only once—I woke up in the middle of the night, wrote something down on a piece of paper, and went back to sleep. Next morning I looked at what I'd scribbled, expecting to see nonsense, and was surprised to see a correct calculation of the crucial proof step I'd been missing.) The risk with new research is always that you may never be able to prove the theorem, and then you may have nothing to show for your years of work. That's why most mathematicians always have at least two research problems going. Putting all your eggs in one basket, like Uncle Petros, and devoting your whole life to a single problem is a very dangerous strategy, especially when that problem is one of the great unsolved puzzles of mathematics.

Goldbach's conjecture, which Uncle Petros studies, was first stated in 1742 by Christian Goldbach: it says that every even number greater than 2 can be written as the sum of two prime numbers; 40, for instance, is 17 + 23. This is a beguilingly simple statement that feels as if it should be easy to prove. But nobody has yet managed to do so. Petros, a brilliant young mathematician, decides at the age of twenty-four that he is the man for the job. In any other field, he says, at that age "he would have been a promising beginner with years and years and years of rich creative opportunities ahead of him. In mathematics, however, he was already at the peak of his powers. He estimated that he had, with luck, at the most ten years in which to dazzle humanity" before his mathematical powers began to wane. This belief means that Petros puts enormous, unsustainable pressure on himself to make rapid progress.

Now I don't buy this "young man's game" business, not from Uncle Petros and not from G. H. Hardy. Admittedly, I have skin in the game here, being neither young nor a man, but even so. Yes, of course some of the most famous mathematicians have done their most important work before the age of forty. But for many of them that's because they did *everything* before the age of forty—it took until the mid-nineteenth century for average life expectancy to exceed that age. It's a romantic notion, but like the cliché that all the best rock stars die at twenty-seven, it doesn't stand up to scrutiny.[5]

The literary mathematicians we've seen so far do not paint a promising picture: the choice seems to be between emotionless logician and tragic prodigy. But there are other ways to be a young mathematician, and you can even be a good detective while you're at it.

The narrator of *The Curious Incident of the Dog in the Night-Time*, by Mark Haddon, is a big fan of Sherlock Holmes. He is Christopher Boone, a fifteen-year-old boy who loves mathematics, seeing it as a consoling oasis of order in a world of chaos. Christopher struggles to understand people's emotions and behaviors, their idioms and impulsiveness. He always tells the truth, because a lie is something that didn't happen. But there are infinitely many things that didn't happen, and once you contemplate lying, you start thinking about them all, and it's overwhelming. The story begins on the night that the neighbor's dog is found dead, and Christopher decides he will solve the mystery of who killed him.

The title of the book is a clue to Christopher's love of Sherlock Holmes: it is a reference to Conan Doyle's short story "The Adventure of Silver Blaze," and to a clever piece of deduction by Holmes, who, like Christopher, sees things that others miss. The story concerns the theft of a champion racehorse, Silver Blaze, and the murder of his trainer. Holmes and

Watson head to Dartmoor to investigate the crime, which they discuss with Inspector Gregory from Scotland Yard. A tallow candle, a milliner's bill, and five gold sovereigns are found on the body of the deceased, and three sheep in a nearby field have gone lame. Inspector Gregory, trying to make sense of the baffling array of evidence, consults Holmes:

> *"Is there any point to which you would wish to draw my attention?"*
> *"To the curious incident of the dog in the night-time."*
> *"The dog did nothing in the night-time."*
> *"That was the curious incident," remarked Sherlock Holmes.*

The fact that the dog did nothing, it's revealed later, is a crucial clue: it shows that the person responsible for the crime was not a stranger—if he were, the dog would have barked.

As Christopher tries to understand what happened to the neighbor's dog, he tells us more about his world and how he navigates it. He has behavioral difficulties, and although no specific label is given in the book, Mark Haddon has since said that if Christopher were diagnosed, it would be with a form of autism. But he stresses that the book is not about a boy with a particular diagnosis; it's about "a young mathematician who has some strange behavioural problems." Haddon made the conscious decision not to do extensive research into the details of autism because, as he says, there is no "typical" person with autism: "They're as large and diverse a group of people as any other group in society." To which I could add: there is no typical mathematician, and they're as large and diverse a group of people as any other group in society.

Christopher talks a lot about mathematics in the book, and prime numbers in particular fascinate him. The chapters of the book are numbered not 1, 2, 3, 4, and so on but 2, 3, 5, 7, 11, and so on—the prime numbers—because Christopher likes them and it's his book. He explains an ancient Greek technique for finding the primes: "First, you write down all the positive whole numbers in the world. Then you take away all the numbers that are multiples of 2. Then you take away all the numbers that are multiples of 3. Then you take away all the numbers that are multiples of 4 and 5 and 6 and 7 and so on. The numbers that are left are the prime numbers." (This is because, if you recall, prime numbers are the ones that are multiples of only themselves and 1, so by eliminating all multiples of smaller numbers, whatever is left is prime.) Christopher expresses the nature of primeness very poetically: "Prime numbers are what is left when you have taken all the patterns away. I think prime numbers are like life. They are very logical but you could never work out the rules, even if you spent all your time thinking about them." I love this description, and I love Haddon's portrayal of Christopher as a fully rounded human being. He is no tragic prodigy. His fully realized character is a counterpoint to the austere mathematical logic of Holmes and Moriarty.

The next mathematician I'd like you to meet offers a delightful contrast to Lisbeth Salander and her unlikely proof of Fermat's Last Theorem. Thomasina Coverly is the exuberant mathematician in Tom Stoppard's joyful play *Arcadia*. The play begins in 1809 with thirteen-year-old Thomasina discussing Fermat's Last Theorem with her tutor, Septimus Hodge (himself no slouch mathematically). He has asked her to prove the theorem, knowing that she will fail, in an attempt to distract her for a while so that he can read poetry in peace. It's a rather pleasing coincidence that *Arcadia* was first performed in 1993, just two months before Andrew Wiles's proof was announced. Since I've not actually told you yet what Fermat's Last Theorem says, perhaps we should hear it from Septimus: "When x, y and z are whole numbers each raised to the power of n, the sum of the first two can never equal the third when n is greater than 2." What does this mean? Well, we all learn Pythagoras's theorem in school. In a right triangle, the square of the hypotenuse is equal to the sum of the squares of the other two sides, right? That means that if we have a right triangle with sides x, y, and z, where z is the hypotenuse—the side opposite the right angle—then it's always the case that $x^2 + y^2 = z^2$.

There are lots of solutions to this that are whole numbers. For example, $3^2 + 4^2 = 5^2$ because $3^2 = 9$, $4^2 = 16$, and $9 + 16 = 25 = 5^2$. Sets of numbers like 3, 4, 5 that satisfy the equation $x^2 + y^2 = z^2$ are called *Pythagorean triples.* I still remember the excitement of discovering when I was Thomasina's age that there are infinitely many whole number solutions (one of the many times my eternally patient mother had to tell me that no, I wasn't the first to notice this). Just take any odd number, square it, and then the whole numbers on either side of half the square go with your odd number to make a triple. So if we take 5, for instance, then 5 squared is 25. Half that is 12½, so the numbers nearest that are 12 and 13. And hey presto, $5^2 + 12^2 = 13^2$. And for 7, we square it to get 49, half of which is 24½, and sure enough, $7^2 + 24^2 = 25^2$. It's such a lovely pattern! (It would seem I'm still excited about it more than thirty years later.)

Given that we have all these examples of solutions to $a^2 + b^2 = c^2$, it shouldn't be too hard to find some examples of $x^3 + y^3 = z^3$, right? Only, no. Nobody could find any. (I should clarify Septimus slightly, because we also ask that the solutions be positive whole numbers, to avoid obtaining boring facts like $0^3 + 0^3 = 0^3$.) The plot thickens. Nobody could find any solutions of $x^4 + y^4 = z^4$ either, or, for that matter, $x^{anything} + y^{anything} = z^{anything}$, when *anything* is any whole number bigger than 2. This was the observation that Fermat made in his marginal note, for which he claimed a marvelous proof.

It's a testament to Stoppard's mathematical literacy that he does not follow the lazy trope of having his genius Thomasina find a proof of Fermat's Last Theorem. Even better, he toys with us by having her say "Oh! I see now! The answer is perfectly obvious." Cue eye rolls from every mathematician in the audience. Septimus replies dryly, "This time you may have overreached yourself," before promising her an extra spoonful of jam with her rice pudding if she can find Fermat's proof. But she responds, "There is no proof, Septimus. The thing that is perfectly obvious is that the note in the margin was a joke to make you all mad."

Thomasina Coverly is a fictional mathematician, but in her portrayal are echoes of a real one who was more or less her contemporary: Ada Lovelace. Ada's mother, Annabelle, was herself mathematically gifted, so much so that she was nicknamed "the princess of parallelograms" by her husband, and Ada's father, Lord Byron. That marriage was by all accounts a complete disaster, and Ada never met her father, who died when she was eight years old. Ada grew up loving mathematics, and in pursuing her studies she became acquainted with many well-known mathematicians and scientists. She is best known now for her work with the mathematician and engineer Charles Babbage on early precursors to the computer. There is at least one computer language named in her honor.

Babbage invented the first mechanical computers, the Difference Engine and the Analytical Engine. At that time, mathematical tables (listing things like logarithms, or sines and cosines) were crucial in navigation and engineering, but they were full of errors, errors that could cost lives. Babbage had the idea of creating computing machines to automate the work. Not all the machines he designed were built, but the ones that have been built have worked. The Analytical Engine has all the characteristics of modern computers—a memory, inputs and outputs, and programmability—the idea was to use punched cards, as the Jacquard looms of that era did. Ada Lovelace worked on the Analytical Engine, creating an algorithm (for finding something called Bernoulli numbers) that has been called the world's first computer program. "We may say most aptly," she wrote, "that the Analytical Engine weaves algebraical patterns just as the Jacquard loom weaves flowers and leaves." She called her approach to mathematics "poetical science."

Babbage perhaps had less of a poetic instinct—certainly his surviving attempts at poetry are pretty dreadful. However, there's a lovely anecdote about an interaction between Babbage and Alfred, Lord Tennyson, that I can't resist sharing. In a 1900 edition of Tennyson's early works, the editor, John Churton Collins, notes that all printed versions of Tennyson's poem "The Vision of Sin" up to 1850 include the lines "Every minute dies a man / Every minute one is born." These lines, says

Collins, caused Babbage to write a tongue-in-cheek letter of complaint to Tennyson:

> *I need hardly to point out to you that this calculation would tend to keep the sum total of the world's population in a state of perpetual equipoise, whereas it is a well-known fact that the said sum total is constantly on the increase. I would therefore take the liberty of suggesting that, in the next edition of your excellent poem, the erroneous calculation to which I refer should be corrected as follows: Every minute dies a man, And one and a sixteenth is born. I may add that the exact figures are 1.167, but something must, of course, be conceded to the laws of metre.*

Collins believes that Tennyson took this objection seriously and substituted "moment," a less precise period of time, for "minute," which finesses away the problem. And that, claims Collins, is why every printing of the poem from 1851 onward reads, "Every moment dies a man / Every moment one is born."

Ada Lovelace spoke of the Analytical Engine weaving algebraical patterns rather than flowers and leaves. But in *Arcadia*, Thomasina tries to bring these ideas together, to work out how we can describe natural phenomena with equations. We've all heard of the "bell curve" in statistics (also known as the "normal distribution"). If there is a curve like a bell, asks Thomasina, why not a curve like a bluebell? She comes up with a great idea for a kind of mathematics that might produce such a curve, and at that point in the play Stoppard can't resist a little in-joke about Fermat. "I, Thomasina Coverly, have found a truly wonderful method whereby all the forms of nature must give up their numerical secrets and draw themselves through numbers alone. This margin being too mean for my purpose, the reader must look elsewhere for the New Geometry of Irregular Forms by Thomasina Coverly."

This "Geometry of Irregular Forms" features shapes produced by repeated iterations, what we now call fractals. If you think about how plants grow, something like a fern emerges from precisely the same sort of process as the dragon curve and snowflake curve that I showed you in the last chapter. Now that we are able to get computers to do the hard work for us (thanks to pioneers like Lovelace and Babbage), we can produce extremely convincing and lifelike images of plants, trees, and other organisms. They have this "self-similarity" so characteristic of fractals, in that zooming into the picture produces something that looks just the same as the original picture. Here's the starting point for a "plant" design I made—just four straight lines:

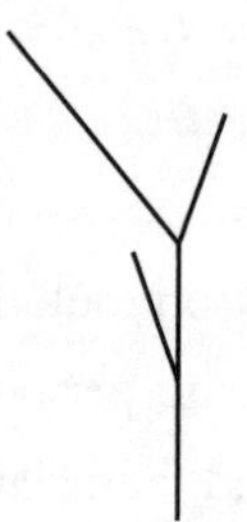

This isn't going to get me hired by Pixar. But then the magic happens. Each iteration adds smaller versions of the initial design at specified points of existing lines, just as a growing fern or tree does. Here's the second iteration—look out for the four smaller copies of the first design:

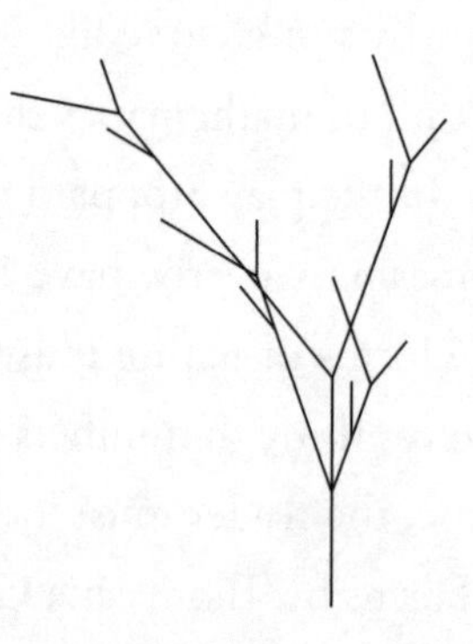

By the sixth iteration, we get something that looks very organic:

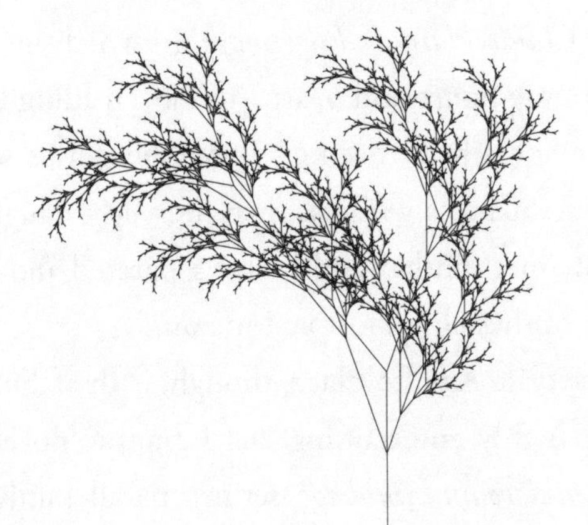

Another example of fractals in the natural world is coastlines—they wiggle in and out at any scale from which they are viewed, and increasing the scale just reveals more of the same structure. River systems also have a fractal structure—as you go upstream, the rivers branch off into smaller rivers, then smaller and smaller streams, each having the distinctive S-shaped curves of the larger waterways they connect to. The same kind of structures can be seen in lightning, with its constantly forking paths, and even in our own bodies—our brains themselves appear to be fractal in design, with bifurcating pathways allowing for maximum connectivity. It really seems that fractals are the geometry of nature. They are, as a line from *Arcadia* has it, "how nature creates itself, on every scale, the snowflake and the snowstorm."

Tom Stoppard has said that Thomasina Coverly is not Ada Lovelace, but many authors have found Lovelace an inspiration. The future British prime minister Benjamin Disraeli wrote a rather overblown novel, *Venetia*, published in 1837, whose title character is a not-very-well-disguised version of Ada. This portrayal focused less on the mathematician than on the racy life of a scandalous poet's daughter. Each age and each writer has their own Ada. The American playwright Romulus Linney wrote *Childe Byron* about the tragedy of the separation of a father and his daughter. He imagines the adult mathematician Ada Lovelace, ill with the cancer that ultimately killed her, grappling with her conflicted feelings about her father. The title *Childe Byron* refers to Byron's most famous poem, *Childe Harold's Pilgrimage*, in which he writes of "Ada! Sole daughter of my house and heart." Linney, reading Byron's lines on Ada, "I see thee not. I hear thee not. But none can be so rapt in thee," felt deeply the resonances with his own life. "My daughter Laura, the actress," he said, "her mother and I were separated and divorced when she was a baby, so these lines just laid me out."

By far my favorite Ada Lovelace, though, is the fabulous heroine of Sydney Padua's highly entertaining 2015 graphic novel, *The Thrilling Adventures of Lovelace and Babbage*, set in a parallel universe where the two protagonists have managed to get the Analytic Engine to work and use it to fight crime and generally be awesome.

The literary mathematicians we've discussed so far are, apart from a few walk-on parts, fictional—even Sydney Padua's "Ada Lovelace" is not meant to be a true-to-life portrayal. But my next example of a mathematician in literature is not just real, she is depicted as such. In *Too Much Happiness*, the Nobel Prize–winning author Alice Munro gives a poignant fictionalized account of the last days in the life of the mathematician Sofya Kovalevskaya.[6] This may be the most human portrayal of a mathematician I have read in literature. Kovalevskaya throughout is not a tortured genius, a freak, an unnatural being. It is certainly part of her story that she struggled to gain acceptance—this was the nineteenth century, after all. But that is not the focus of the writing. She has some unhappiness in her personal life, but this is not lazily portrayed as a necessary consequence of being a mathematician. That she didn't end up in a long-lasting, loving relationship is not because she is a cold-hearted logician unable to interact with human beings. She isn't solving differential equations because she can't get a husband, nor is she failing to get a husband because she insists on solving differential equations. In Munro's tale, as in life, sometimes these things just happen, or don't.

In the story, we follow Kovalevskaya on her return to Stockholm University from a visit to her mathematician friend and mentor Karl

Weierstrass, who had made her the first, and at that time only, female mathematics professor in Europe. Munro, as is her right, plays slightly with the time lines, but her narrative fits well with the known facts, and she gives us a convincing version of Kovalevskaya, lauded by the mathematical community in France after winning one of its most prestigious prizes, but still an outsider. "They had given her the Bordin Prize, they had kissed her hand and presented her with speeches and flowers in the most elegant lavishly lit rooms. But they had closed their doors when it came to giving her a job. They would no more think of that than employing a learned chimpanzee." Thankfully the world has moved on, but there are some resonances for me, as a woman mathematician, even though more than a century separates the start of Kovalevskaya's working life and the start of mine. Even now, the myth persists in some circles that women are in some way not as suited to mathematics as men.

Of course, I was not the first woman to study at my university, as Kovalevskaya was at hers, Heidelberg University in Germany, where she registered in 1869. Still, my former Oxford college, Balliol, had been admitting women students for only fourteen years (after more than seven hundred years as an all-male institution) when I arrived there as an undergraduate in 1993. There are still places in the world where girls are not even allowed to attend school. Munro alludes to these challenges without ever being heavy-handed. Her writing has a lovely economy—she has Kovalevskaya's lover remark that she should perhaps return to Sweden because her students and her daughter need her—"a jab there, a suggestion familiar to her, of faulty motherhood?" I identified quite hard with that one! When I returned to my mathematics lectureship after my first maternity leave, a colleague asked, "Can your husband not afford to let you give up work?"—the assumption being that I must surely want to stop. The colleague in question had children himself. I suppose his wife couldn't afford to let him stop doing mathematics, poor thing, so he had to struggle on.

Kovalevskaya grew up in a well-to-do Russian family that, while wishing their daughters to be educated to a certain degree (presumably enough to secure eligible husbands), frowned upon the young Sofya's unseemly enthusiasm for mathematics. Perhaps, though, her fate was written on the wall—literally. In her autobiography, *A Russian Childhood*, she recalled that when her family moved to the country, they ran out of wallpaper partway through decorating the nursery and just used random paper to finish the job. But "by happy chance, the paper for this preparatory covering consisted of the lithographed lectures of Professor Ostrogradsky on differential and integral calculus, which my father had acquired as a young man." Sofya spent many hours staring at this "mysterious wall," trying to parse its curious sentences. Years later, at fifteen, she started learning calculus, and her tutor was amazed at how quickly she picked up the concepts, almost as if she had known them in advance. "And, as a matter of fact, at the moment when he was explaining these concepts I suddenly had a vivid memory of all this, written on the memorable sheets of Ostrogradsky, and the concept of limit appeared to me as an old friend."

Kovalevskaya's mathematical success was hard-won. At that time it was impossible for Russian women to study at Russian universities, and unmarried women could leave the country only with their father's permission. Sofya's father would never have agreed, so, as Munro's story recounts, she entered into a "white marriage" with a young man sympathetic to the cause. Vladimir Kovalevsky, a paleontologist, married Sofya in Russia; they then traveled together to Germany and lived apart while they both pursued their studies. Several years later they did end up in a sexual relationship (we're all human), and Sofya had a child by him, but she and Vladimir soon became estranged again, and he eventually committed suicide. Though this was tragic, it may have helped Kovalevskaya's career, because a widow is (or was at that time) much more respectable than an estranged wife.

Alice Munro, with her great skill as a short story writer, paints for us, in a few beautifully crafted vignettes, a rich and compelling portrait of Kovalevskaya. The unfolding tragedy of her early death is interspersed with her reflections and memories of her struggles to balance mathematics with everything else she wanted from life. The triumph of being the first woman to gain a mathematics Ph.D. is succeeded by the temptation to rest on her laurels. "She was learning, quite late, what many people around her appeared to have known since childhood—that life can be perfectly satisfying without major achievements. It could be brim full of occupations that did not weary you to the bone." For a while, she used her talents in ways "not so disturbing to other people or so exhausting to herself, as mathematics." But mathematics, her old friend, was waiting for her when she was ready to return. In fact, after Vladimir's death, she refused to eat for five days but then seems to have resolved that life must go on, and that mathematics could provide a solace. "She asked for paper and pencil," writes Munro, "that she might continue working on a problem."

Kovalevskaya's mathematics was deep and important. Weierstrass said that each of the three papers she presented for her thesis was worthy of a Ph.D. in its own right. The work she did for the Bordin Prize represented a major advance in a problem of classical mechanics that had been studied by Euler and Lagrange. She was also a writer. As a teenager she knew Dostoyevsky—in fact she had something of a crush on him, so it must have been a bit of a blow when he proposed marriage to her older sister (it didn't happen—their father disapproved). She also met George Eliot at a literary salon on a trip to England. As well as her autobiography, which was well received (one enthusiastic contemporary reviewer even compared it to Tolstoy's *Childhood*), she published a novel, *Nihilist Girl*, as well as plays, poetry, and short stories. She had several works in progress at the time of her death. Who knows what she might have achieved had she lived longer.

Readers of this book will, I hope, be convinced already that there is nothing unnatural about combining mathematics and literature. But Kovalevskaya had this to say to a friend who questioned it: "Many people who have never had occasion to learn what mathematics is confuse it with arithmetic and consider it a dry and arid science. In actual fact it is the science which demands the utmost imagination." She continues:

> *It is impossible to be a mathematician without being a poet in soul. . . . One must repudiate the old prejudice by which poets are supposed to fabricate what does not exist, and that imagination is the same as "making things up." It seems to me that the poet must see what others do not see, must see more deeply than other people. And the mathematician must do the same.*

Mathematics is an important part of Sofya Kovalevskaya's life, but Alice Munro does not let it define Kovalevskaya as a human being—that's one reason why *Too Much Happiness* is such a great piece of writing. The same is true for the fictional mathematician who is a central character in Chimamanda Ngozi Adichie's magnificent *Half of a Yellow Sun*. The book tells the story of the devastating 1967–70 Nigerian-Biafran war through the eyes of a handful of people caught up in it. The war was an unimaginable tragedy. It's estimated that more than a million people died. Adichie's novel is a stunning, compassionate story of those years. I urge you to read it. Much of the story involves Odenigbo, a professor of mathematics at Nsukka University; his eventual wife, Olanna; and their live-in servant, the houseboy Ugwu.

In Odenigbo, Adichie, like Alice Munro, creates not a stereotype but a totally believable person. He is charismatic, idealistic, flawed. He is passionate about politics, about his Igbo identity, and about education. As he says, "How can we resist exploitation if we don't have the tools to understand exploitation?" Don't get me wrong, he loves mathematics too. When Olanna moves to Nsukka to live with Odenigbo, it does not stop him from heading off the very next day for a mathematics conference. Even that, though, is important not just for mathematics but for personal reasons: "He would not have gone if the conference was not focused on the work of his mentor, the black American mathematician David Blackwell. 'He is the greatest living mathematician, the greatest,' he said."

I remember hearing an academic, thankfully retired now, claiming that it's not possible to introduce cultural diversity into the math curriculum.

The "reason" he gave was that Black mathematicians are such a recent phenomenon that any of their work would be too advanced to explain to his undergraduate students. Utter nonsense, of course—I'd only that week been telling my first-year students about research on magic squares by the Nigerian mathematician Muhammad Ibn Muhammad Al-Fulani Al-Kishwani (who died in 1741). It was even worse when I remembered that the course this colleague was teaching was on what's known as game theory, one of whose most important figures is the selfsame David Blackwell who mentored Odenigbo.

The word "trailblazer" is perhaps overused, but if anyone has earned it, it's Blackwell. After earning his Ph.D. at the University of Illinois in 1941 at only twenty-two years of age—the seventh Ph.D. in mathematics ever awarded to an African American—he then did a year's fellowship at Princeton's Institute for Advanced Study but was prohibited from attending classes or doing research at Princeton University, which at the time did not admit Black students and had no Black faculty, despite the university's collaborative relationship with IAS. Blackwell would go on to chair the Statistics Department at the University of California, Berkeley, for thirty years. But the first time he applied for a post there he was turned down because the wife of the head of the Mathematics Department, whose official role included hosting dinners for faculty members, refused to entertain the idea of having a person of color in her house.

Over the course of his career, David Blackwell published more than eighty academic papers and was highly influential in mathematics, with dozens of Ph.D. students, well-regarded textbooks, and a reputation as an excellent teacher. What's he doing in Adichie's book, though? Well, Adichie was brought up in Nsukka after the Biafran war. Her mother was an academic registrar, and her father, James Nwoye Adichie, was, guess what, professor of statistics at Nsukka University. I wouldn't be so reductive as to say that Odenigbo "is" Adichie's father—he certainly isn't—but there are interesting crosscurrents in their stories. James Nwoye Adichie earned his Ph.D. at UC Berkeley while David Blackwell was department chair, and he would certainly have known him. Blackwell wasn't his direct supervisor, but I've checked the records, and he did supervise at least two other Nigerian Ph.D. students. I like to think that the young Chimamanda heard her father speak of him.

I also had a look at James Adichie's publications. It gave me pause to see a telltale gap between 1967 and 1974. That bland absence, so easy to overlook, but what turmoil and trauma it conceals. In *Half of a Yellow Sun*, when Olanna and Odenigbo return to their old home after the war, they find that most of their books and papers have been burned. Odenigbo "began to search through the charred paper, muttering, 'My research papers are all here, *nekene nke*, this is the one on my rank tests for signal detection . . .'" This little detail is poignant. We can't know what papers Adichie senior might have published between 1967 and 1974 if the war hadn't happened, but such a title would fit nicely with his real-life paper "Rank Tests in Linear Models." As Odenigbo and his family start to rebuild their lives after the war, "books came for Odenigbo from overseas. *For a war-robbed colleague*, the notes read, *from fellow admirers of David Blackwell in the brotherhood of mathematicians*."

Chimamanda Ngozi Adichie has talked about the problem of the "single story," in the context of stereotypes of "Africans." She recounted how "a student told me that it was such a shame that Nigerian men were physical abusers like the father character in my novel [*Purple Hibiscus*]. I told him that I had just read a novel called *American Psycho* and that it was such a shame that young Americans were serial murderers." Of course she does not actually think this, because she, and all of us, have been exposed to many stories of America and what it means to be American. The danger of having only a single story, a single version of an American, or a Nigerian, or, dare I say it, a mathematician, is that a single story creates stereotypes, and the problem with stereotypes, says Adichie, "is not that they are untrue, but that they are incomplete." In literature, as in life, there are as many different ways to be a mathematician as there are different ways to be a person.

Acknowledgments

This is my first book, and I have been blessed with an amazing team supporting me every step of the way. Jenny Heller, my agent, has taken me from a vague thought of writing a book through to publication in the space of two years, and has been incredible. It has been a revelation to work with my editor, Caroline Bleeke, and her assistant, Sydney Jeon, who have made the book so much better in so many ways. Bob Miller and the whole Flatiron team have been fantastic, and have helped to make the daunting process of writing a book as smooth as possible.

I'm grateful to my head of department at Birkbeck, Ken Hori, and the dean of my school, Geoff Walters, for letting me take a sabbatical term in autumn 2021 to focus on the book, and to Maura Paterson and Steve Noble, my colleagues and friends in the mathematics group, without whom I wouldn't have made it through the pandemic with even the modicum of sanity I have now.

The team at Gresham College have been brilliant in supporting me as Gresham Professor of Geometry. My program of lectures, focusing as it has on the links between mathematics and the arts and humanities, has quite a different flavor from some previous series, and I'm grateful that they supported my vision and let me run with it.

It was my appointment as Gresham Professor that led to Siobhan Roberts approaching me for a profile in *The New York Times*, and that article opened a lot of doors for me, so, Siobhan, I definitely owe you a good dinner next time you are in London!

Thanks also to Sir Ian Livingstone for giving his time so generously for me to pick his brains about how he writes his epic *Fighting Fantasy* books.

I am lucky to have a wonderful group of friends who have shared this journey (and many others) with me. Thank you to Caroline Turner, who among other things introduced me to Charlotte Robertson and Jenny Heller at the fabulous Robertson Murray Literary Agency. Thank you to Rachel Lampard, who reads the Booker Prize short list with me every year, and has been so kind and supportive over the last few years. Thank you to Alex Bell, whom I met when we were very pregnant with our first babies, and our families have been close ever since. Alex, you're a rock! I'm sorry I couldn't sneak the words supercalifragilisticexpialidocious, hippopotomonstrosesquipedalian, floccinaucinihilipilification, honorificabilitudinitatibus, contraremonstrance, and epistemophilia into the book like you dared me to. Or could I? And thank you, too, to my wonderful book club, the Ladies Wot Read. We've been meeting every month since 2006 and seen one another through good and bad times with love and support (and occasionally we also discuss the book). Alex, Claire, Claire, Colette, Emma, Hadassah, Lucy, and Rachel: thank you.

I have the great good fortune to have grown up in a house filled with both books and ideas. My dad, Martin, trained me and my sister, Mary, up very well as independent researchers by sending us to a dictionary every time we asked the meaning of a word. Mary put up with my many foibles, and she didn't mind coming with her little sister to the geological museum during my rocks-and-minerals phase, or talking with me on long car journeys about four-dimensional shapes during my . . . well, that phase is still ongoing, to be fair. My mum, Pat, who died in 2002 of multiple sclerosis, was always there for me when I was young, supplying cuddles, soothing away my (many) worries, and of course asking me interesting mathematical questions when I was bored. She went with me to lectures by the then–Gresham Professor of Geometry Christopher Zeeman. I wish she could have known that one day it would be her little girl up there. I think of you every day, Mum. Thank you.

My brilliant, beautiful daughters, Millie and Emma, bring so much joy into my life. (They have also taught me the importance of being at peace with chaos.) They've had so much to deal with in the past few years, and have coped amazingly. You can't write a book without time, and they have given me that time with a good grace.

Finally, and most important, my husband, Mark. I find it hard to put into words how much he has done for me. He has always given me unconditional support for everything I have wanted to do, and this book is no exception. I'm the family worrier, and he's the chief morale officer. He talks me down from my periodic "why on earth did I think I could do this" freak-outs. He makes me a nice cup of tea when he can sense I need one. I know he will always be there for me, and I for him. He's the best husband and father it's possible to be. This book wouldn't exist without him.

Notes

1: One, Two, Buckle My Shoe

1. The zeros are represented by ten-letter words, in case you were wondering! If you want to compose a bit of pilish of your own, the first forty digits are 3.141592653589793238462643383279502884197.
2. Pound says that "an 'image' is that which presents an intellectual and emotional complex in an instant of time. . . . It is this presentation of such a "complex" instantaneously which gives that sense of sudden liberation; that sense of freedom from time limits and space limits; that sense of sudden growth, which we experience in the presence of the greatest works of art" ("A Few Don'ts by an Imagist," *Poetry*, Chicago, March 1913). He explained that just as in poetry, the same mathematical expression can often be interpreted on several different levels.
3. For a comprehensive academic treatment, try *The Poetics of Japanese Verse—Imagery, Structure, Meter*, by Koji Kawamoto (University of Tokyo Press, 2000). I'd also recommend Abigail Friedman's *The Haiku Apprentice: Memoirs of Writing Poetry in Japan* (Stone Bridge Press, 2006), a charming account of her experiences learning to write *haiku* while working as an American diplomat in Tokyo. In terms of online resources, an excellent place to start is the website www.graceguts.com, run by the poet and *haiku* expert Michael Dylan Welch.
4. A full set of all fifty-two Genji-ko diagrams, as well as some from *The Arte of English Poesie*, is presented in the essay "Two Thousand Years of Combinatorics" by the mathematician and computer scientist Donald Knuth, which appears in *Combinatorics: Ancient & Modern*, edited by Robin Wilson and John J. Watkins and published by Oxford University Press in 2013. He believes, like me, that

the best way for humans to communicate, whether mathematics or anything else, is through story. This extends to his philosophy of computer programming, which he says would be much improved if we considered programs to be works of literature. He also wrote a novel in 1974. *Surreal Numbers: How Two Ex-students Turned On to Pure Mathematics and Found Total Happiness* is, as perhaps the title indicates, rather of its time. But it deserves a mention in this book because it is, to my knowledge, the only novel to introduce a piece of mathematics research—the invention of new kinds of numbers by the mathematician John Conway—before it was published anywhere else.

5. Copyright for this important 2021 poem rests with the author, Emma Hart, aged ten at the time of composition. It is reproduced here with her gracious permission.
6. There are several translations of the Temple Hymns, but this one, by Sarah Glaz, is my favorite. Glaz is a respected mathematician and poet whose published books include both an academic textbook on abstract algebra and a book of poetry, *Ode to Numbers* (Antrim House, 2017). The title takes its name from a poem by the Chilean poet Pablo Neruda.

2: The Geometry of Narrative

1. From a public lecture given by Vonnegut in 2004 at Case Western Reserve University. You can watch it online at https://youtu.be/4_RUgnC1lm8.
2. The quote is from Lockwood's 2021 novel, *No One Is Talking About This.*
3. Towles was interviewed for the BBC Radio 4 Bookclub on April 8, 2021. At the time of writing, the episode is available on the BBC iPlayer at https://www.bbc.co.uk/programmes/m000tvgy.
4. The spiraling effect is converging around a point that turns out to be exactly two-thirds of the way along the square and one-third of the way up (if you're a mathematician you might enjoy proving this).
5. Perec made this remark in his article "Four Figures for *Life: A User's Manual*," which appears in English translation in the anthology *Oulipo Compendium* (Atlas Press, 2005). It is noted there that the girl in question appears on pages 231 and 318 in the English edition.

3: A Workshop for Potential Literature

1. There is a translation by Ian Monk of *Les Revenentes* with the title *The Exeter Text: Jewels, Secrets, Sex.* It comes out with a lipogrammatic difficulty level of $0.25398 \times 36,000 = 9,143$, though again this is not a fair comparison once you take into account the superadded challenge of translation.

4: Let Me Count the Ways

1. Not everyone approved of the *Fighting Fantasy* books. A church group published an eight-page pamphlet with dire warnings about the dangers of reading them, saying that because you are interacting with ghouls and demons, you might get possessed by the Devil: "A worried housewife in deepest suburbia phoned her local radio station and said that having read one of our books her child levitated." This didn't seem to put people off. "Kids are thinking—what, for £1.50 we can fly? I'll have some of that!" Teachers, on the other hand, were delighted that these books were actually getting kids reading. It's reported that they increased literacy by 20 percent, and it's certainly good for the vocabulary—"Hey, Dad, what's a sarcophagus?"
2. If you'd like to see another example of a reverse poem, I highly recommend Brian Bilston's "Refugees," which he has made available online.
3. Nesting a narrative like this is sometimes known as *metalepsis*, by the way, for any lovers of ancient Greek out there. There's an example of many-layered metalepsis in another of the stories from *Lost in the Funhouse.* In "Menelaiad," there are a full seven nested stories. Menelaus (king of Sparta) struggles to find his way through the labyrinth of his own narrative: "When will I reach my goal through its cloaks of story?" he asks in desperation.
4. Coe's biography of B. S. Johnson is excellent—it's the source of much of the biographical information I include here and is well worth a read. Jonathan Coe, *Like a Fiery Elephant* (Picador, 2004).
5. I can't say "sonnets" because I'm showing off with an *e*-free sentence.

5: Fairy-Tale Figures

1. Suppose we start a "say what you see" sequence at 42 (after all, according to *The Hitchhiker's Guide to the Galaxy,* 42 is the answer to "Life, the Universe, and Everything"). We'd get a sequence starting 42, 1412, 11141112, 31143112. The brilliant mathematician John Conway, subject of a wonderful 2015 biography by the Canadian author Siobhan Roberts, studied "say what you see" sequences and found they have some truly remarkable properties, which are worth googling if you ever want your mind blown by how much amazing math can come from a little puzzle.
2. I'm certain that there are, but that's just a hunch—nobody has a mathematical proof yet.
3. Bonus alternative version for mathematicians: There are 10 kinds of people in the world. Those who understand binary, those who don't, and those who weren't expecting this joke to be in base 3.

4. The essay is "The Number Three in American Culture," in *Every Man His Way: Readings in Cultural Anthropology* (Prentice-Hall, 1968). By all accounts, Dundes, an American folklorist, was not afraid of controversy—he even received death threats after he wrote an article called "Into the Endzone for a Touchdown," positing a homoerotic subtext in the language and rituals of American football.

6: Ahab's Arithmetic

1. I go into a bit more detail in my paper "Ahab's Arithmetic: The Mathematics of Moby-Dick," *Journal of Humanistic Mathematics* 11, no. 1 (January 2021): 4–32, https://scholarship.claremont.edu/jhm/vol11/iss1/3, DOI: 10.5642/jhummath.202101.03.
2. Blaise Pascal is maybe best known outside mathematics for what's now known as *Pascal's wager*, which essentially says that humans, in our behavior, are betting for or against the existence of God. There are four possibilities: you believe and God exists; you believe and God doesn't exist; you don't believe but God exists; and you don't believe and God doesn't exist. If you believe and God exists, then great (assuming you behave accordingly). Off you go to Heaven for all eternity. If you believe and you are wrong—there is no God—then you lose out on perhaps some pleasures during your finite life, perhaps people laugh at you, you have to get up early to go to church, and so on. But your losses are finite. On the other hand, suppose you don't believe. If there is no God, then again, you're okay. But if there is a God, then you will go to Hell for all eternity, so your losses are infinite. Even if you think the probability that God exists is tiny, there is still a nonzero probability. Anything nonzero multiplied by infinity is infinity. Therefore, if you act purely rationally, says Pascal, you should act as if God exists and try to believe in God, because the expected gains of believing are infinite (however low the probability of God's existence), whereas the expected losses of not believing are also infinite.
3. George Eliot had strong views about the education of women, and *The Mill on the Floss* (1860) hints at them. Brother and sister Maggie and Tom Tulliver have very different educational experiences. Euclid is wasted on Tom, who doesn't get along with geometry at all, but Maggie, who could have gained great pleasure from it, is not given the opportunity. Later on, she begins teaching herself, from Tom's Euclid and some of his other schoolbooks. She "began to nibble at the thick-rinded fruit of the tree of knowledge, filling her vacant hours with Latin, geometry and the forms of syllogism, and feeling a gleam of triumph now and then that her understanding was quite equal to these peculiarly masculine studies."

4. Derek Ball's dissertation, *Mathematics in George Eliot's Novels*, is at the time of writing available from the University of Leicester, in the UK. You can download it from https://leicester.figshare.com/articles/thesis/Mathematics_in_George_Eliot_s_Novels/10239446/1. If you are interested in the broader links between mathematics, science, and creativity in the nineteenth century, Professor Alice Jenkins of the University of Glasgow has written extensively on these topics. Her book, *Space and the "March of Mind": Literature and the Physical Sciences in Britain, 1815–1850* (Oxford University Press, 2007), is an academic exploration of the conversation between science and literature in nineteenth-century Britain.
5. Can I tell you a secret? I haven't read *Finnegans Wake*; I've just dipped in and out. I felt much better about this when I read a brilliant article called "*Finnegans Wake* for Dummies," by Sebastian D. G. Knowles, which I highly recommend if you can get hold of the fall 2008 issue of the *James Joyce Quarterly*. The first sentence is this: "I begin with a confession: in September 2003, after attending two decades of Joyce symposia, teaching over a dozen courses on Joyce, writing a book entirely devoted to Joyce's work, and editing another, I had still not yet read *Finnegans Wake*." In desperation, he signed up to teach a course on it, and that's what finally did the trick.
6. In Chapter 3, I stated the fifth postulate of Euclid in a different way: that given a line and a point not on the line, there is exactly one line through that point, parallel to the given line. This version, which is known as *Playfair's axiom*, after the Scottish mathematician John Playfair, who publicized it in the eighteenth century, is logically equivalent to the one Joyce is referencing, but much easier to work with. It's the version that Hilbert used in his *Foundations of Geometry* that we mentioned in Chapter 3. Joyce's version is what was in the original Greek text.

7: Travels in Fabulous Realms

1. The tallest person who has ever lived is Robert Wadlow (1918–1940). He had a pituitary gland disorder, which meant that his body produced too much growth hormone, and this condition continued throughout his life. By the time he was eight, he was taller than his father; when he died, aged twenty-two, he was 8 feet 11.1 inches (2.72 meters) tall and weighed 315 pounds (199 kilograms). He had a job with the International Shoe Company doing promotional work—the schtick was that if they could make the size 37 shoes that Wadlow wore, then they could make shoes for anyone. Wadlow required leg braces when walking and had little feeling in his legs and feet—ultimately this was what caused his death, because he didn't feel that a minor abrasion

from one of his leg braces had become infected until it was too late to stop the sepsis from spreading.

2. Some editions have "algebraists" rather than "geometers."
3. The essay appeared in the book *Possible Worlds and Other Essays*, published by Chatto and Windus in 1927, but is easy to find online too. Haldane also has bad news about angels during a discussion of flight in which he explains that if you scale up something like a bird by a factor of four, the power required for it to fly increases by a factor of 128. He goes on to say that an angel "whose muscles developed no more power, weight for weight, than those of an eagle or a pigeon would require a breast projecting for about four feet to house the muscles engaged in working its wings, while to economize in weight, its legs would have to be reduced to mere stilts."
4. There have been a couple of real scientific papers on the plausibility of tiny people like Lilliputians and Borrowers, which make for entertaining reading if you like that sort of thing (and who doesn't?). I didn't want to make the calculations of Lilliputians' recommended calorie intake even more complicated, but a 2019 paper suggests that Lilliputians actually need 57 calories, not the 9.3 in my rough estimate, because of Quetelet's observations about how mass changes with height. But that just makes things even worse in terms of Lilliput's economy. Check out T. Kuroki, "Physiological Essay on *Gulliver's Travels*: A Correction After Three Centuries," in *The Journal of Physiological Sciences* 69 (2019): 421–24. Meanwhile, in "What Would the World Be Like to a Borrower?" (*Journal of Interdisciplinary Science Topics* 5, 2016), J. G. Panuelos and L. H. Green give more detail about several aspects of life for Borrowers, including discussions of their voices—likely too high-pitched and faint for us to hear them.

8: Taking an Idea for a Walk

1. Just as any picture of a three-dimensional cube cannot render every side as a square, any drawing of a hypercube necessarily distorts some of the lengths. One way to get around the challenge of representing a three-dimensional cube on a two-dimensional page is to draw what's called the *net* of the cube. This is a diagram with six squares that can be cut out and folded up in three dimensions to make a cube. In the same way, we can make a three-dimensional net of eight cubes that could be folded up in four dimensions to make a hypercube. It's this version of the hypercube that appears in Salvador Dalí's famous 1954 painting *Crucifixion (Corpus Hypercubus)*.
2. These strange new geometries seemed incomprehensibly alien to many. Even Ivan Karamazov, the most intellectual of the Karamazov brothers in Dostoyevsky's

1880 novel, struggled with them. Ivan compares trying to understand the divine to trying to understand non-Euclidean geometry. There are geometers and philosophers, he says, who "dare to dream that two parallel lines, which according to Euclid can never meet on earth, may meet somewhere in infinity. I have come to the conclusion that, since I can't understand even that, I can't expect to understand about God. I acknowledge humbly that I have no faculty for settling such questions, I have a Euclidean, earthly mind, and how could I solve problems that are not of this world?"

3. *Flatland* wasn't the first time that two-dimensional beings trying to grasp a three-dimensional world had been invoked as an analogy. The German physicist Hermann von Helmholtz had discussed how we would interpret the world if we were two-dimensional beings living on the surface of a sphere. The most important precursor to *Flatland* is the article "What Is the Fourth Dimension?" by Charles Howard Hinton, which was almost certainly read by Abbott. Hinton was a mathematician, teacher, and writer, a great popularizer of science. The article asks us to imagine "a being confined to a plane" and to suppose "some figure, such as a circle or rectangle, to be endowed with the power of perception." Does that sound familiar?
4. Angles in triangles on the surface of a sphere really do add up to more than 180°. We don't have to worry about it in everyday life, because it turns out that the amount by which the sum exceeds 180° is proportional to how much of the sphere's surface is contained in the triangle. But this knowledge was vital for one of the more impressive technical feats of the nineteenth century: the Great Trigonometrical Survey, a seventy-year undertaking with the aim of mapping the whole of India with high precision. The distances being measured were so vast that the curvature of the earth, and its effect on the angles in a triangle, had to be taken into account in the calculations.
5. The story is "The Unparalleled Adventure of One Hans Pfaall."
6. "Cryptographic mathematicians," the book says, are "by nature high-strung workaholics." Are they, though? Anecdote is not data, but I met some of the mathematicians at Royal Holloway (where Brown's other cryptographer heroine, Sophie Neveu, is supposed to have trained) when I gave an invited seminar there a few years ago, and they were a lovely bunch who gave every appearance, over tea and biscuits, of being relaxed and genial, and not at all addicted to workahol.
7. Okay fine, RSA stands for Rivest-Shamir-Adleman. It's really not their fault that they get all the credit and fame. They earned it. Clifford Cocks doesn't begrudge them it either. As he said, "You don't get into this business for public recognition." You can read more about RSA in Simon Singh's excellent history of cryptography, *The Code Book*, which includes this Clifford Cocks quote in its chapter on public key cryptography.

9: The Real Life of Pi

1. *Life of Pi* is far from being the only work of literature to mention this fascinating number. For maximum erudition, check out this passage from the beginning of Umberto Eco's *Foucault's Pendulum* (a novel I've heard described as "the thinking person's *Da Vinci Code*"): "That was when I saw the Pendulum. . . . I knew—but anyone could have sensed it in the magic of that serene breathing—that the period was governed by the square root of the length of the wire and by π, that number which, however irrational to sublunary minds, through a higher rationality binds the circumference and diameter of all possible circles. The time it took the sphere to swing from end to end was determined by an arcane conspiracy between the most timeless of measures: the singularity of the point of suspension, the duality of the plane's dimensions, the triadic beginning of π, the secret quadratic nature of the root, and the unnumbered perfection of the circle itself." You don't get *that* in Dan Brown.
2. My quotations are from James E. Irby's English translation of Jorge Luis Borges, *Labyrinths*, Penguin Classics edition (2000).

10: Moriarty Was a Mathematician

1. Fermat's Last Theorem has been rolled out in this way on several occasions over the years. Pre-Wiles, your characters could gain fame and fortune by finding any proof. Post-Wiles, they have to find the elusive "short" proof. There was an episode of the British TV show *Doctor Who* in 2010 in which the Doctor tells the "real proof"—in other words, the short one—of Fermat's Last Theorem to a group of geniuses as proof that they should trust his intelligence. By contrast, Jorge Luis Borges, with his customary erudition, carefully doesn't claim that Unwin, the mathematician in "Ibn Hakkan al-Bokhari, Dead in His Labyrinth," has proved the theorem. He has simply published a paper on "the theory supposed to have been written by Pierre Fermat in a page of Diophantus." There's another twist on it in the 1954 short story "The Devil and Simon Flagg," by Arthur Porges, in which the mathematician Simon Flagg tricks the Devil into a wager. If the Devil can answer one question, he can have Flagg's soul. If not, he must give Flagg wealth, health, and happiness all his days and leave him in peace for eternity. The question: "Is Fermat's Last Theorem true?" The Devil fails to answer, and Flagg gets his reward.
2. Beth is good at math, as it happens—Tevis tells us that she is at the top of her class at the orphanage. That's actually a crucial part of the story, because it means that she is the one chosen to go down to the basement after the Tuesday arithmetic class to clean the board erasers—something considered a privilege. It's

there that she first sees the janitor playing chess and persuades him to teach her. I'm pleased to report that Tevis does not, as the TV adaptation did, give Beth's mother a backstory as a suicidal mathematician.

3. Hardy also made a major contribution to mathematics by taking the time to read a letter he received from a complete stranger one day, a clerk from India with no formal mathematical education, that was filled with crazy-looking formulae like $1 + 2 + 3 + 4 + \cdots = -\frac{1}{12}$. There is a context in which this formula makes sense, and Hardy recognized that whoever had written this letter, and derived this mathematics more or less by pure intuition, had a rare talent. He managed to get funding to bring his correspondent, whose name was Srinivasa Ramanujan, over to England to work with him. Ramanujan proved to be one of the most brilliantly original mathematical thinkers of the twentieth century, and it is to Hardy's credit that he recognized his gift and did all in his power to support him. Ramanujan's story is told in the wonderful 2007 play *A Disappearing Number*, by Simon McBurney and his theater company, Complicité.
4. It shows even more in Doxiadis's bestselling 2009 graphic novel *Logicomix*, co-written with the computer scientist Christos Papadimitriou, about the twentieth-century quest for the foundations of mathematical truth, which is also highly recommended. At the start of the twentieth century, there was a concerted attempt to put the whole of mathematics on the strictest possible logical foundation. The idea was to try to create a sort of mathematical language in which every possible mathematical statement could be expressed. Then you could agree on a list of initial assumptions, or axioms, and, proceeding in accordance with strictly defined rules of inference, either prove or disprove each statement. But the mathematician Kurt Gödel blew the whole thing out of the water in 1931 when he proved that any such mathematical system must be inadequate in that there would be true statements you could make that could not be proved within the system. This development was a profound disappointment to the logicians who had made it their life's work to try to systematize the whole of mathematics. Perhaps that's why the *New York Times* review of *Logicomix* was titled "Algorithm and Blues."
5. For example, Jimi Hendrix, Kurt Cobain, Janis Joplin, Jim Morrison, Amy Winehouse, and Brian Jones died at twenty-seven. But Elvis Presley, John Lennon, David Bowie, and hundreds of others did not.
6. There is some ambiguity about how best to transliterate Софья Васильевна Ковалевская into the English alphabet. Russian full names have three parts, the first name, the patronymic (based on the father's first name), and the surname. So because Sofya's father's name was Vasily, and her husband's surname was Kovalevsky, her full name was Sofya Vasilyevna Kovalevskaya. Both the patronymic and the surname have a male and a female form. Quite often, people

are addressed by their first name and patronymic, and, to add to the fun, lots of first names have diminutive forms. Anyone who's ever read a Russian novel knows the problem—you read ten increasingly confusing pages about this Sasha chap who's suddenly appeared before you realize it was just Aleksandr Petrovich all along. Anyway, with Sofya Kovalevskaya I've gone for what current consensus seems to believe is the most accurate rendering. But you'll also see Sofia, Sophia, Sophie, or even the diminutive Sonya, as well as Kovalevsky, Kovalevski, Kovalevskaia, and even Kovalevskaja. Alice Munro went for Sophia Kovalevsky.

A Mathematician's Bookcase

We have come to the end of our mathematical guided tour of the house of literature. Mathematics is there in the foundation, in the rhythms of poetry and the structures of prose. It is there in the decoration of the house, the metaphors and allusions. And it is there in the characters who move through the house, who bring it to life.

I've gathered here a collection of some of the books on my shelves that we have discussed—with a few bonus recommendations thrown in for good measure. I hope that I have given you a new perspective on both mathematics and literature, and new ways to enjoy them both. May this be just the start of your journey. Happy reading!

1: One, Two, Buckle My Shoe

Tom Chivers (editor), *Adventures in Form: A Compendium of Poetic Forms, Rules and Constraints* (Penned in the Margins, 2012).

Jordan Ellenberg, *Shape: The Hidden Geometry of Absolutely Everything* (Penguin Press, 2021). He has also written a well-received novel, *The Grasshopper King* (Coffee House Press, 2003).

Michael Keith's pilish poem "Near a Raven" is available on his website, cadaeic.net (that strange word "cadaeic" is not in any dictionary—but if you let $a = 1$, $b = 2$, and so on, you will see what's going on). He has also written an entire pilish book, the only one I know of: Michael Keith, *Not a Wake: A Dream Embodying (Pi)'s Digits Fully for 10000 Decimals* (Vinculum Press, 2010).

Raymond Queneau's *Cent mille milliards de poèmes* has been translated into English more than once. Stanley Chapman's version uses the rhyme scheme *abab, cdcd, efef, gg,* and Queneau's reaction was apparently "admiring stupefaction," so that seems a good place to start. It appears in *Oulipo Compendium*, edited by Harry Mathews and Alastair Brotchie (Atlas Press, 2005).

Murasaki Shikibu, *The Tale of Genji,* translated by Royall Tyler (Penguin Classics, 2002).

For poetry with explicitly mathematical themes, check out these three collections:

Madhur Anand, *A New Index for Predicting Catastrophes* (McClelland and Stewart, 2015).

Sarah Glaz, *Ode to Numbers* (Antrim House, 2017).

Brian McCabe, *Zero* (Polygon, 2009).

2: The Geometry of Narrative

Eleanor Catton, *The Luminaries* (Little, Brown, 2013).

Georges Perec, *Life: A User's Manual,* translated by David Bellos (Collins Harvill, 1987).

Hilbert Schenck, "The Geometry of Narrative," *Analog Science Fiction / Science Fact* (Davis Publications, August 1983).

Catherine Shaw has written several Vanessa Duncan novels. The first is *The Three-Body Problem* (Allison and Busby, 2004).

Laurence Sterne, *The Life and Opinions of Tristram Shandy, Gentleman* (1759–1767).

Amor Towles, *A Gentleman in Moscow* (Viking, 2016).

3: A Workshop for Potential Literature

Christian Bök, *Eunoia* (Coach House Books, 2001).

Alastair Brotchie (editor), *Oulipo Laboratory: Texts from the Bibliothèque Oulipienne* (Atlas Anti-Classics, 1995).

Italo Calvino, *If on a Winter's Night a Traveler,* translated by William Weaver (Harcourt Brace Jovanovich, 1982).

Italo Calvino, *Invisible Cities,* translated by William Weaver (Harcourt Brace Jovanovich, 1978).

Mark Dunn, *Ella Minnow Pea: A Novel in Letters* (Anchor, 2002).

Harry Mathews and Alastair Brotchie (editors), *Oulipo Compendium* (Atlas Press, 2005).

Warren F. Motte, *Oulipo: A Primer of Potential Literature* (Dalkey Archive Press, 1986).

Georges Perec, *A Void*, translated by Gilbert Adair (Harvill Press, 1994).

Georges Perec, *Three by Perec*, translated by Ian Monk (David R. Godine Publisher, 2007). This contains *The Exeter Text: Jewels, Secrets, Sex*, Monk's translation of *Les Revenentes*, the novel that prohibits every vowel except *e*.

4: Let Me Count the Ways

John Barth, *Lost in the Funhouse*, reissue edition (Anchor, 1988).

Julio Cortázar, *Hopscotch, Blow-Up, We Love Glenda So Much* (Everyman's Library, 2017)—this contains *Hopscotch* as well as a collection of short stories including "Continuity of Parks."

B. S. Johnson, *House Mother Normal*, reissue edition (New Directions, 2016).

B. S. Johnson, *The Unfortunates*, reissue edition (New Directions, 2009).

Gabriel Josipovici, *Mobius the Stripper* (Gollancz, 1974).

Ian Livingstone and Steve Jackson wrote many of the "You are the hero" books in the *Fighting Fantasy* series. The ones I mentioned are *The Warlock of Firetop Mountain* (Puffin, 1982), co-written with Steve Jackson, and *Deathtrap Dungeon* (Puffin, 1984), both of which were reissued by Scholastic Books in 2017.

US readers may remember the Choose Your Own Adventure series, which had its heyday in the 1980s. Most of the books were written by Edward Packard or R. A. Montgomery. I'm pretty sure I had *The Abominable Snowman* (Bantam Books, 1982).

5: Fairy-Tale Figures

Annemarie Schimmel's book *The Mystery of Numbers* (Oxford University Press, 1993) devotes a chapter not quite to each number (she'd be writing it for all infinity if she did) but to all the small numbers. It's in this little book that I first learned that cats have different numbers of lives depending on their nationality.

If you'd prefer a purely mathematical guide to numbers and their properties, you can't go wrong with *The Penguin Dictionary of Curious and Interesting Numbers* by David Wells (Penguin, 1997).

For a really deep dive into the language of numbers and the origin of number words and number symbols in different languages and cultures, try Karl Menninger's *Number Words and Number Symbols: A Cultural History of Numbers* (Dover, 1992). It has quite an old-fashioned tone of voice (it's a translation of the 1958 German edition) but is full of fascinating nuggets.

6: Ahab's Arithmetic

Herman Melville, *Moby-Dick* (1851).

George Eliot's novels all contain mathematical allusions. We discussed *Adam Bede* (1859), *Silas Marner* (1861), *Middlemarch* (1871–1872), and *Daniel Deronda* (1876).

Vasily Grossman, *Life and Fate* (NYRB Classics, 2008).

Leo Tolstoy, *War and Peace* (1869).

James Joyce, *Dubliners* (1914), and *Ulysses* (1922). I'm not going to tell you to read *Finnegans Wake* (1939).

7: Travels in Fabulous Realms

Mary Norton, *The Borrowers* (1952). There were several later Borrowers books too.

François Rabelais, *Life of Gargantua and Pantagruel* (published in English 1693–1694).

Jonathan Swift, *Gulliver's Travels* (1726).

Voltaire, *Micromégas* (1752).

8: Taking an Idea for a Walk

Books relating to *Flatland* and the fourth dimension

Edwin A. Abbott, *Flatland: A Romance of Many Dimensions* (1884). You might also like Ian Stewart's *The Annotated Flatland* (Perseus Books, 2008).

Dionys Burger, *Sphereland* (Apollo Editions, 1965).

A. K. Dewdney, *The Planiverse: Computer Contact with a Two-Dimensional World* (Poseidon Press, 1984).

Fyodor Dostoyevsky, *The Brothers Karamazov* (1880).

Charles H. Hinton, *An Episode of Flatland: or, How a Plane Folk Discovered the Third Dimension* (S. Sonnenschein, 1907).

Rudy Rucker, *The Fourth Dimension and How to Get There* (Penguin, 1986).

Rudy Rucker, *Spaceland: A Novel of the Fourth Dimension* (Tor Books, 2002).

Ian Stewart, *Flatterland* (Perseus Books, 2001).

Books relating to fractals

Michael Crichton, *Jurassic Park* (Arrow Books, 1991).

John Updike, *Roger's Version* (Knopf, 1986).

Several Richard Powers novels discuss fractals, including *The Gold Bug Variations* (Harper, 1991), *Galatea 2.2* (Harper, 1995), and *Plowing the Dark* (Farrar, Straus and Giroux, 2000), in which an artist works with computer scientists to design a virtual world using, in part, fractals.

Books relating to cryptography

Dan Brown, *The Da Vinci Code* (Doubleday, 2003) and *Digital Fortress* (St. Martin's Press, 1998).

Arthur Conan Doyle, "The Adventure of the Dancing Men" (in *The Return of Sherlock Holmes*, 1905) and *The Valley of Fear* (1915). Both had previously appeared in *The Strand Magazine*.

John F. Dooley (editor), *Codes and Villains and Mystery* (Amazon, 2016), is an anthology that includes the O. Henry story "Calloway's Code."

Robert Harris, *Enigma* (Hutchinson, 1995).

Edgar Allan Poe, "The Gold-Bug" (1843) and "The Purloined Letter" (1844); both available in numerous short story collections and editions of Poe's works.

Neal Stephenson, *Cryptonomicon* (Avon, 1999).

Jules Verne, *Journey to the Center of the Earth* (1864).

Hugh Whitemore, *Breaking the Code* (Samuel French, 1987).

9: The Real Life of Pi

Jorge Luis Borges, *Labyrinths*, Penguin Modern Classics edition (Penguin Books, 2000). This collection includes "The Library of Babel" as well as several other wonderful stories with a mathematical flavor. "The Library of Babel" is also included in William G. Bloch, *The Unimaginable Mathematics of Borges' Library of Babel* (Oxford University Press, 2008).

Lewis Carroll, *Alice's Adventures in Wonderland* (1865) and *Through the Looking-Glass, and What Alice Found There* (1871). For mathematical discussions of Lewis Carroll's work, I recommend Martin Gardner's *The Annotated Alice* (Penguin Books, 2001) and Robin Wilson's *Lewis Carroll in Numberland* (Penguin Books, 2009).

Yann Martel, *Life of Pi* (Mariner Books, 2002).

10: Moriarty Was a Mathematician

Chimamanda Ngozi Adichie, *Half of a Yellow Sun* (Knopf, 2006).

Isaac Asimov, *Foundation* (Gnome Press, 1951), the first of seven books in the *Foundation* series.

Apostolos Doxiadis, *Uncle Petros and Goldbach's Conjecture* (Faber and Faber, 2001).

Mark Haddon, *The Curious Incident of the Dog in the Night-Time* (Doubleday, 2003).

Aldous Huxley, "Young Archimedes" (1924). It is the first story included in Clifton Fadiman's *Fantasia Mathematica* (Simon and Schuster, 1958). This anthology contains a broad selection of mathematically themed short stories, poetry, and quotations. I have to say that some of them haven't aged particularly well, but the collection is still worth dipping into.

Sofya Kovalevskaya, *A Russian Childhood* (Springer, 1978) and *Nihilist Girl* (Modern Language Association of America, 2001).

Stieg Larsson, *The Girl Who Played with Fire* (Knopf, 2009)—the second in the *Millennium* series, following *The Girl with the Dragon Tattoo*.

Alice Munro, *Too Much Happiness* (Knopf, 2009).

Sydney Padua, *The Thrilling Adventures of Lovelace and Babbage: The (Mostly) True Story of the First Computer* (Pantheon Books, 2015).

Tom Stoppard, *Arcadia: A Play in Two Acts* (Faber and Faber, 1993). I also recommend his play *Rosencrantz and Guildenstern Are Dead* (Faber and Faber, 1967), for its fascinating exploration of probability, chance, and fate.

Walter Tevis, *The Queen's Gambit* (Random House, 1983).

There are many books featuring mathematicians that we didn't have the space to discuss. Here are a few to get you started:

Catherine Chung, *The Tenth Muse* (Ecco, 2019), the story of a brilliant young mathematician taking on the Riemann hypothesis, one of the great unsolved problems of mathematics. It weaves in stories of real women mathematicians who, in Chung's words, "posed as schoolboys, married tutors, and moved across continents, all to study and excel at mathematics."

Apostolos Doxiadis and Christos Papadimitriou, *Logicomix: An Epic Search for Truth* (Bloomsbury, 2009), a graphic novel narrated by a fictional Bertrand Russell and featuring mathematicians such as David Hilbert, Kurt Gödel, and Alan Turing.

Jonathan Levi, *Septimania* (Overlook Press, 2016). This entertaining novel features mathematician Louiza, along with Isaac Newton and Newton expert Malory, about whom Levi says that "in the Kingdom of Mathematicians, he as a Historian of Science . . . bestrode the River Cam with the charisma and stature of a Colossus." As 2021–23 president of the British Society for the History of Mathematics, I can confirm that all our members are incredibly charismatic and you will probably become more so yourself if you join.

Simon McBurney / Théâtre de Complicité, *A Disappearing Number* (Oberon, 2008), a stage play about the Indian mathematician Srinivasa Ramanujan and his work with G. H. Hardy.

Yoko Ogawa, *The Housekeeper and the Professor* (Picador, 2009), a touching and poignant story of a mathematics professor who lives with only eighty minutes of short-term memory, and the friendship that develops with his housekeeper and her son.

Alex Pavesi, *Eight Detectives* (Henry Holt, 2020). This novel centers on a mathematician who has analyzed the permutations of murder mysteries. I don't want to tell you anything about it because I'll spoil it for you, but I very much enjoyed it.

Index

About the Author

Sarah Hart is a respected pure mathematician and a gifted expositor of mathematics. Since 2013, she has been a professor of mathematics at Birkbeck, University of London. Educated at Oxford and Manchester, Dr. Hart is also the thirty-third Gresham Professor of Geometry, the first woman to hold the position since its inception in 1597.

Recommend

Once Upon a Prime

for Your Next Book Club!

Reading Group Guide available at

flatironbooks.com/reading-group-guides